George Anderson

The universe is intelligent. The soul exists.

Quantum mysteries, multiverse, entanglement, synchronicity. Beyond materiality, for a spiritual vision of the cosmos.

Copyright 2019
Bruno Del Medico Publisher
Sabaudia (LT) Italy
Communications: edizioni@delmedico.it
Only Italian or English, please. Other languages will be ignored.

HOW TO GET COPIES OF THIS PRINTED BOOK OR E-BOOK

Publisher's secure site: https://www.qbook.it
The whole catalog. Reduced prices. 20 languages

Book index

Book index 5
Introduction 10
Living in the shell of a walnut 13
    What does Hamlet have with Stephen Hawking? 14
    An author who represents his time. 15
    Tycho Brahe and the supernova N1572.... 21
Astronomical disputes 27
    The Ptolemaic system 28
    The Copernican revolution 30
    Tycho Brahe and the tonic model 35
    Thomas Digges and the heliocentric model 37
"De l'infinito, universo e mondi" 43
    Giordano Bruno, the philosopher of the infinite 44
    Giordano Bruno and his idea of infinity ... 50
Ethics issues 55
    The problems of an infinite universe 57
Infinity in a finite space 61

A premonitory dream 62
The mandala 67
Jung and the Mandalas 72
The cosmic egg 83
The cosmic egg and current physics 90
The meeting between Jung and Pauli 91
The psychic diagram of Pauli and Jung ... 95
Let's talk more about Mandala 103
The infinite in the finite 103
Thought is a "finite" that contains the infinite 110
Cosmic Thought 112

The multiverse theory 116

The theory of the multiverse 118
Quantum physics is the mother of the multiverse 120
First phase. A slit 125
Second phase. Two slits 126
Third phase. The role of the observer 126
How many types of multiverse exist? 132
The multiverse landscape 132
The quantum multiverse 133
The simulated multiverse 134
The ultimate multiverse 135
The brane multiverse 136
Aesthetics of science 136

Intelligence at the center of the universe 139

The role of the observer 140

- A scientific nemesis 143
- Amazing coincidences 147
- The anthropic principle 157
- Birth and evolution of the anthropic principle 160
- Is man really at the center of the universe? 167
- Intelligence cooperation 171

Creatio ab nihilo 179

- What evidence do we have on the intelligence of the "Cosmic Matrix"? 180
- But does the universe made of matter really exist? 184

Non locality, entanglement............................ 189

- Einstein and the locality 190
- Is causality the basis of all things? 192
- Quantum entanglement 195
- Everything is one in the non-local dimension 198

The soul exists 201

- The aggregation of matter 203
- The matter aggregates into coherent and finalized forms 206
- Every aggregation comes from a project 208
- Quantum atmospheres 209
- What makes us conscious? 213
- Quantum physics and the soul 214
- Collapse of quantum waves 220

- *Neurons as qubits*......... 221
- *Angels, demons and souls of the dead*.... 224
- *Collective unconscious and archetypes*.. 227
- *The strange coincidences*......... 230
- *Take the challenge*......... 237
- *Meditation and prayer*......... 239

Appendix 1. Hamlet......... 245
- *Characters*......... 246
- *The plot of the tragedy*......... 247

Glossary......... 251
Bibliography......... 267

Introduction

The incredible discoveries of quantum physics are completely upsetting the assumptions of classical science. Today the technique allows amazing achievements. For example, the first quantum computers with almost unlimited computing capabilities are being realized. Some support the real possibility of time travel. In addition to these innovations known to the general public, there are others less known but no less important. They are the novelties deriving from quantum studies, among which we can mention the "superposition of states" and the "quantum collapse".

The "superposition of states" confirms that the same particle can be found simultaneously in two or more places. The theory of "quantum collapse" confirms that the behavior of matter can be decided simply by observation. These are not assumptions, but principles verified experimentally.

This book does not only deal with these innovations, but gives much space to more advanced theories. These are theories announced but not yet confirmed. Furthermore, the book also evaluates the most risky theories, provided they are scientifically based.

For example, the book talks about the multiverse, or theory of parallel universes, proposed by the physicist Hugh Everett. In the same way the book speaks of non-locality. It is a psychic space totally independent of the laws of classical physics. As a result of non-locality, elementary particles, located at astronomical distances, behave as if they were one.

This book also talks about the latest research by Roger Penrose, an unbelieving physicist, and Stuart Hameroff. According to these two scientists the soul exists and can be identified with quantum fluctuations. These fluctuations have the ability to survive the physical death of the body.

If really the "souls" are condensations of quantum fluctuations, we can formulate a question: will it ever be possible to devise instruments that allow dialogue with these fluctuations?

The book exposes the research of established scientists but without using any mathematical formula. The theories are exposed in a simple and understandable way to everyone. In this way everyone can discover the unsuspected aspects of the reality in which we live.

It is clear that quantum physics is decreeing the end of materialism and the beginning of a new cultural phase, based on the collaboration between spirit and matter.

Living in the shell of a walnut

*My goal is simple. It is the complete understanding of the universe.
I want to understand why the universe is made as it is and why it actually exists.*

(Stephen Hawking, astrophysicist)

What does Hamlet have with Stephen Hawking?

On March 14, 2018, in Cambridge, one of the most famous scientists of our time, astrophysicist Stephen Hawking, passed away. His interests ranged over vast areas of knowledge. For example, he first carried out scientific studies on the astronomical alignments of Stonehenge.

Hawking was also a very valuable communicator. His most famous work, the book "A Brief History of Time" was published in 1988 and has sold more than ten million copies worldwide.

In 2001 Hawking released another hit, "The Universe in a Nutshell". The title is quite original so as not to arouse curiosity. In fact, the reference to the nut shell is not explained in the introduction. We can only find a reference to the beginning of the third chapter. Here is a quote from Shakespeare's Hamlet:

> "O God, I could be bound in a nutshell and count myself a king of infinite space."
> *(Hamlet, Act II)*

Hawking is a man of vast culture. There is a specific reason why he decided to choose this sentence. This chapter will be intended to explain the reason for this choice. Starting from this quote we will be able to understand the topics discussed below.

An author who represents his time.

William Shakespeare lived between 1564 and 1616. He produced many works. Among these the most famous is certainly the Hamlet, which Shakespeare wrote between 1600 and 1602.

Hamlet is a tragedy and tells some apparently fantastic events that, however, can be linked to the political and cultural context of the time. Indeed, Shakespeare fills the narrative with overtones. Consequently the content of the work can be evaluated on different levels. We can distinguish the narrative level and the historical level. But there is also a third level. This can be considered as the transposition of the author's opinions in relation to the cultural ferment of the time.

The three levels are summarized in the attached diagram. The knowledge of the detailed plot of Hamlet is not essential for understanding the different interpretations. We can briefly recall that one of the main characters is King Claudius. Claudio married Gertrude, the widow of the late King Hamlet. Incidentally, Gertrude is the mother of Prince Hamlet. (Curiously, the young prince has the same name as his father).

The ghost of the deceased king appears to his son Hamlet and reveals the secret of his death. He claims to have been killed by Claudio. Claudio committed the crime to usurp the kingdom and marry Gertrude. Anyone who wants can find a summary of the plot in the Appendix.

In the narration the vicissitudes of a negative character, the king Claudius, murderer and usurper, intertwine with those of a victim, Prince Hamlet. The prince, while being right, must pretend to be mad to avoid other negative actions of Cladio. Shakespeare adopts the "good-bad" scheme because he derives it from his cultural connections and the scientific disputes that involve him. He attributes the role of the "villain", represented by King Claudius, to the astronomer Tycho Brahe.

The universe is intelligent. The soul exists.

Figure 1 - Stephen Hawking on the day of his first marriage to Jane Wilde, which occurred in 1963. Shortly thereafter he was struck by the illness that forced him into a wheelchair for life.

Instead, the role of the "good" is interpreted by Prince Hamlet. In the underlying plot of Shakespeare, however, the "good" is another astronomer, Thomas Digges.

Evidently, the two astronomers supported different theories and Shakespeare decidedly sided with one of the two, namely for Digges.

This means that Sakespaere sided with the Copernican thesis on the position of the Earth in the universe, supported by Digges. This thesis was opposed to that supported by Brahe, of Ptolemaic orientation.

The Copernican thesis foresaw that the Earth revolved around the Sun, while the Ptolemaic one foresaw, on the contrary, that the Sun turned around the Earth.

Interpretive levels of Shakespeare's Hamlet

King Claudio
Narrative level. The ghost of Hamlet's father reveals that Claudio killed him to steal the throne and marry the widowed queen Gertrude.

Historical level. Claudius is identified with Frederick II (1534-1588) who was king of Danimara and Norway. When the astronomer Tycho Brahe becomes famous, Claudio gives him an island located near the castle of Elsinore.

Allusive level. King Claudius represents the Ptolemaic thesis, supported by Tycho Brahe. This thesis places the Earth at the center of the universe. Shakespeare does not approve of this thesis.

Rosencrantz and Guildenstern
Narrative level. They are friends of Hamlet. Claudio summons them and assigns them the task of investigating Hamlet's madness.

Historical level. Two practically equal surnames appear among the ancestors of Tycho Brahe.

Allusive level. The two characters accept the task of convincing Hamlet but they cannot. They represent traditional science that continues to support Ptolemy's thesis but is about to be supplanted by the Copernicus thesis.

Queen Gertrude
Narrative level. Wife of the late King Hamlet. Immediately after the death of her husband she agreed to marry Claudio.

Historical level. Gertrude is Queen Sofia, wife of Frederick II and mother of Christian IV.

Allusive level. Probably there was a romantic relationship between Sofia and Tycho. This binds Tycho's Ptolemaic theses even more to the prevailing establishment in that historical period.

Bernardo

Narrative level. In the first act of the tragedy, Bernard mentions a star that appeared in the sky and brought misfortune.

Historical level. That star would be the supernova that appeared in the skies of Europe in 1572 and described by Tychio Brahe.

Allusive level. The new star announces misfortune because it coincides with the appearance of the ghost of the late King Hamlet.

Prince Hamlet

Narrative level. The ghost of his father reveals the crime committed by Claudio.

Historical level. Prince Hamlet is identified with King Christian IV. When he ascends to the throne, Christian begins a trial against Tycho Brahe and forces him to emigrate to Prague. Cristiano probably wants revenge on Tycho's relationship with his mother Sofia.

Allusive level. Hamlet represents the thesis supported by Thomas Digges. Digges supports the model of the universe proposed by Copernicus. In this model the Sun is at the center of the universe.

Copernicus diminishes the role of the Earth which is no longer at the center of everything. But Hamlet does not consider this downgrading important. He feels happy even living in a Nutshell

Tycho Brahe and the supernova N1572

Tycho Brahe was a brilliant young man. In 1572, at the age of 27, he gained international fame by describing the explosion of a supernova. Today we identify this astronomical event with the initials N1572 or with the name "Eta-Cassiopeiae B", "the supernova of Tycho".

In the eyes of the uninitiated, a supernova resembles a new, very luminous star that suddenly appears in the sky. In Tycho's time the appearance of new celestial objects was a source of great concern, because this phenomenon was interpreted as a baleful omen. In fact, Shakespeare puts this event right at the beginning of Hamlet, as if to herald the tragedy of the events narrated below.

In the first act of Hamlet Bernardo, a military man at the king's service, arrives in the castle's stands to give Francesco a change of guard. Shortly afterwards, Marcello and Orazio also arrive. These four characters speak of the apparitions of the specter of the king, who died two months earlier. The apparitions are associated with the path in the sky of the new star. Bernardo relates the facts in this way:

> "Last night of all, When yond same
> star that's westward from the pole.

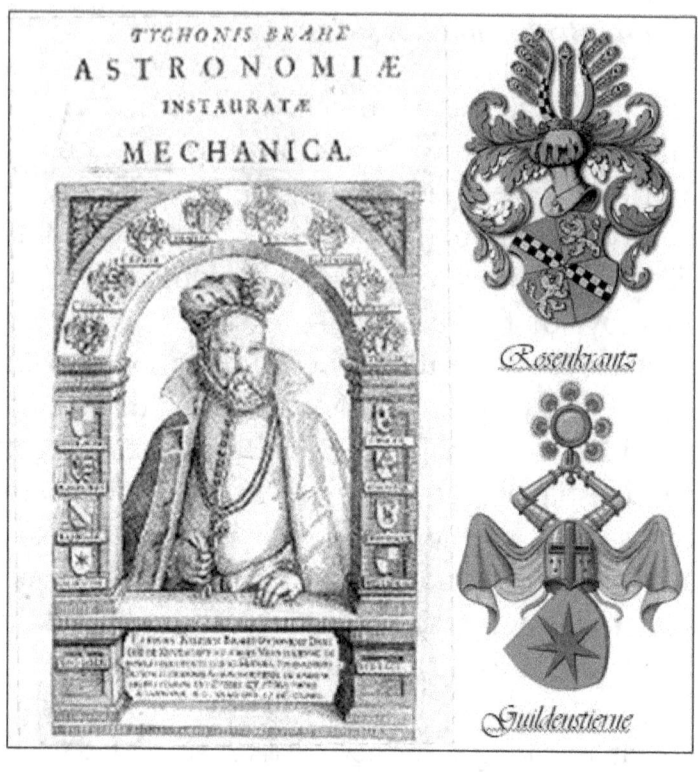

Figure 2 - Portrait of Tycho Brahe surrounded by the coats of arms of his ancestors, two of which bear the surnames of Rosenkrantz and Guildenstierne. These surnames are extraordinarily similar to those of two characters who were Hamlet's fellow students.

The universe is intelligent. The soul exists.

> Had made his course t' illume that part of heaven. Where now it burns, Marcellus and myself, The bell then beating one..."

But at that very moment, along with the star, the specter of the king also appears.

When the supenova explosion occurred in 1572, Shakespeare was eight years old. Surely the event impressed him a lot and was relevant in his cultural growth.

Tycho Brahe also observed the phenomenon on the evening of 11 November 1572:

> "Suddenly and unexpectedly I saw an unknown star in the zenith, with a very bright light."

The supernova had a brightness comparable to that of the planet Venus. It was also visible in the sky by day. Tycho described the phenomenon in a small volume published in 1573 with the title "*De nova stella*".

The supernova ceased to shine in 1574 but, metaphorically, Tycho's good star began to shine from that very moment. The astronomer became so famous internationally that King Frederick II (in the tragedy, Claudius) gave him the island of Hven located near his castle of Elsinore, at the entrance to the Øresund strait. On this island Tycho had a castle built which he named Uranienborg, in honor of the muse of astronomy, Urania.

The story ends in an unedifying way. It seems that Tycho has become the lover of Queen Gertrude, widow of the killed king. In the historical reality Gertrude was Queen Sofia, wife of Frederick II and mother of his successor Christian IV.

Obviously Cristiano IV did not like the astronomer's relationship with his mother. When he ascended the throne, the new king decisively changed the relationship between the royal house and Tycho and began a judicial process against him.

Following this, in 1597 Tycho left the island of Hven and emigrated to Prague. Both the Uranienborg castle and the nearby astronomical complex of Stjerneborg were destroyed shortly after the astronomer's death. In the 1950s, archaeological excavations were carried out at Stjerneborg. Later the site was rebuilt. Currently Uranienborg houses a museum dedicated to Tycho Brahe and the history of the island of Hven.

As for the knowledge of the supernova, until the last century no one knew what kind of celestial object it was. After 1952 astronomers began to study the emissions of the sky in the radio frequency band. This made it possible to identify the remains of Tycho's supernova with the object 3C10. It seems that this supernova was generated by the explosion of a white dwarf that had crossed the limit of Chandrasekhar, sucking up matter from another star. In 2005 the astronauts also identified the other star of the binary system and called it Tycho G.

Astronomical disputes

Not from the stars do I my judgement pluck;
And yet methinks I have Astronomy,
But not to tell of good or evil luck,
Of plagues, of dearths, or seasons' quality;
Nor can I fortune to brief minutes tell,
Pointing to each his thunder, rain and wind,
Or say with princes if it shall go well
By oft predict that I in heaven find:
But from thine eyes my knowledge I derive,
And, constant stars, in them I read such art
As truth and beauty shall together thrive,
If from thyself, to store thou wouldst convert;
Or else of thee this I prognosticate:
Thy end is truth's and beauty's doom and date.
(William Shakespeare, Sonnet XIV)

The Ptolemaic system

The dispute in these pages is restricted to the thought of Tycho Brahe and Thomas Digges. However, before going into the matter it is appropriate to briefly set out the question that gave rise to the dispute. The two had different beliefs about the shape and functioning of the universe. In this regard, many theories were elaborated throughout human history.

The Greeks were the first to make a model of the solar system. Hipparchus, a philosopher who lived between 200 and 120 BC, carefully studied the observations and knowledge accumulated over the centuries by the Babylonian Chaldeans. Ipparco used this knowledge to develop a model able to explain the motion of the Sun and that of the Moon.

In the 2nd century AD the model developed by Claudius Ptolemy established itself. Ptolemy was a Greek of Hellenistic language and culture. He was an astrologer, astronomer and geographer. He lived in Alexandria of Egypt between 100 and 175 AD (*figure 3*).

Ptolemy proposed the so-called Ptolemaic or geocentric model. According to this model, the solar system is a large sphere placed at the center of the Universe. The Earth is flat and immobile, and is located in the center of the celestial sphere. the Sun, the Moon and the other planets revolve around the Earth

Finally, Ptolemy claims that the boundary of the univero consists of the sphere of fixed stars. According to Ptolemy the universe is full and has borders, so it is limited in space. Ptolemy's universe is not infinite.

In the Middle Ages Ptolemy's model was still accepted, but with different interpretations. There were two main interpretations.

An interpretation was called "mathematical astronomy" and was founded on Ptolemy's principal work, "*Almagesto*". This iterpretation was suitable for making calculations and forecasts but was not very organic.

The second interpretation, called "physical cosmology" was based on the work "*De Caelo*" by Aristotle. This interpretation was anthropocentric and was logically consistent. Unfortunately, he could not explain some physical phenomena so it was not consistent on a practical level.

During the Middle Ages the Church supported the Ptolemaic system through scholastic philosophy. In fact, this system is illustrated by Dante Alighieri in the *Divine Comedy*.

The Copernican revolution

The Copernican revolution begins in 1543. In this year Mikołaj Kopernik publishes the "*De revolutionibus orbium coelestium*" (The revolutions of the celestial bodies).

Mikołaj Kopernik was a Polish astronomer, born in Toruń on 19 February 1473 and died in Frombork on 24 May 1543 (figure 4).

His name was Italianized as Niccolò Copernico.

Copernicus replaced Ptolemy's geocentric system with a heliocentric system. While Ptolemy placed the Earth at the center of the universe, Copernicus established that in the center was the Sun.

Even for Copernicus the universe is full and has borders. But at the center of everything there is the Sun. The Earth is not immobile but revolves around the Sun.

It should be noted that the theory of Copernicus is inspired by the heliocentrism of Aristarchus, a Greek astronomer who lived in Samos between 310 and 230 BC. about. Thus, Copernicus was not the first to sustain the centrality of the Sun. But Copernicus was the first to demonstrate the centrality of the Sun with mathematical procedures.

The book of Copernicus, "De revolutionibus", initially had a scarce diffusion even among the experts, that is in the mathematical and astronomical environments of the time. Someone judged the book contemptuously. This bias has continued until recently. In 1959 the philosopher Arthur Koestler wrote his work "I sleepwalkers" in which he talks about the book "De revolutionibus" and defines it "... the book that no one has ever read".

For many decades the Copernicus theory was substantially ignored by the establishment of the time. Only very few university courses cited the Copernican theory together with the Ptolemaic theory, which was taught regularly.

The Catholic Church was not initially hostile: the "De revolutionibus" was considered by Jesuit astronomers and mathematicians. These, however, decidedly preferred the "ticonic" system, developed by Tycho Brahe between 1587 and 1588.

However we must recognize that in 1582, some Copernicus calculations were used in the reform of the Gregorian calendar.

The "De revolutionibus" was inserted by the Holy Office in the "Index librorum prohibitorum" or "Index of the forbidden books". Fortunately, this happened a few decades after publication. This delay allowed Copernicus' theory to spread even the religious cultural environments most opposed to his idea.

The definitive statement of the heliocentric theory is due to the fathers of modern astronomy, Galileo Galilei and Isaac Newton.

Figure 3 - Ptolemy was a Greek astronomer and geographer who lived in Alexandria of Egypt between 100 and 175 AD He proposed the so-called Ptolemaic or geocentric system.

The universe is intelligent. The soul exists.

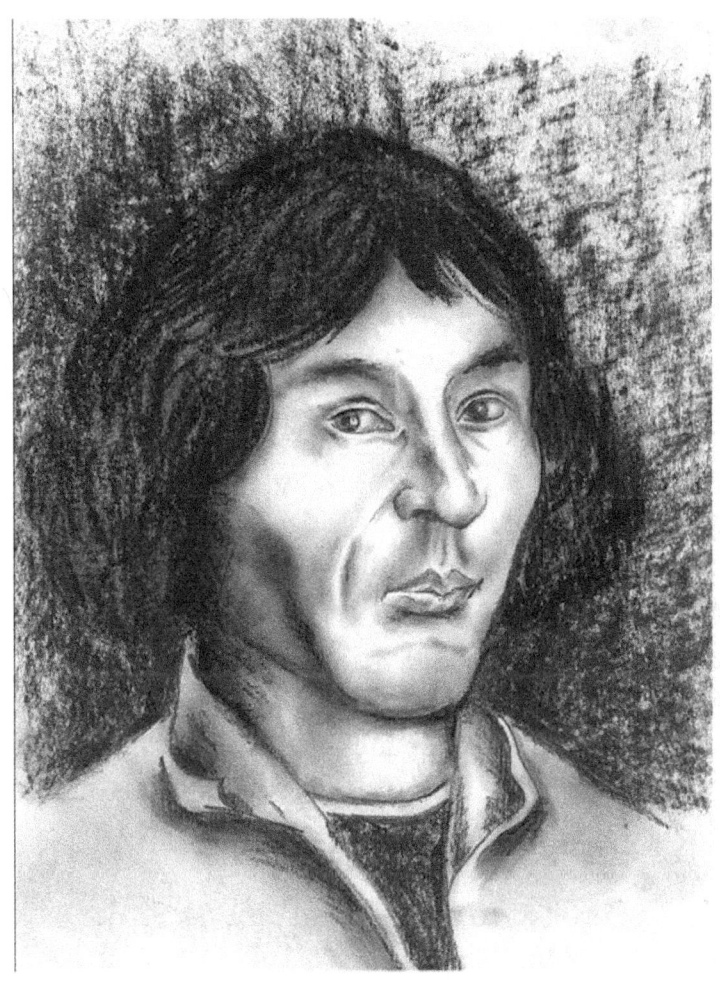

Figure 4 - Niccolò Copernico, Polish astronomer and astrologer, replaced Ptolemy's geocentric system with a heliocentric system.

Before Galileo, Tycho Brahe had suggested a compromise between the Ptolemaic and the Copernican model by proposing the so-called "ticonic" model. According to this model all the planets rotate around the Sun. However, the Sun and the Moon rotate around the Earth. But Galileo rejected this theory.

The importance of Copernicus was recognized in England before other places, thanks above all to Thomas Digges, who supported the Copernican model in his essay "A Perfection Description of the Caelestial Orbes".

Another contribution was given by the publication of Giordano Bruno's book "La cena delle ceneri"" published in 1584 in London by John Charlewood.

In the conclusion of the "De revolutonibus" Copernicus exposes seven points that summarize his theory. I remember someone:

- The orbits and the celestial spheres do not have a single center.
- The center of the Earth is not the center of the Universe.
- All celestial spheres rotate around the Sun. Thus, the center of the Universe is located near the Sun.

- The distance between the Earth and the height of the firmament does not make the movements of the fixed stars perceptible.
- Whatever movement appears in the firmament does not depend on the firmament itself, but on the Earth. The Earth revolves around its poles. The firmament, however, remains immobile.

Later, some of these points will be more correctly specified by Kepler and other studies.

Compared to the previous cosmological models, the Copernican model had a great revolutionary significance. The philosopher Immanuel Kant first coined the term "Copernican revolution". This literary term is still used today, in a figurative sense, to indicate processes of reversal of the fundamental paradigms oof an argument.

Tycho Brahe and the tonic model

Tycho Brahe cultivated his passion for astronomy since adolescence. He studied the texts of antiquity, in particular the "*Almagesto*" of Ptolemy and the "*De revolutionibus*" of Copernicus. However, he did not share either of the two hypotheses about the position of the planets.

In fact, the Danish astronomer represents a turning point between the concepts of ancient and modern astronomy. In 1588 Tycho published "*De mundi aetherei recentioribus phaenomenis*" in which he disputed the Ptolemaic system according to which everything revolves around the Earth. He imagines a hybrid situation, a system known as the "Ticonian system. According to this system the planets rotate around the Sun, but the same Sun with the other planets rotates around the Earth. The earth remains immobile in the center of the cosmos. (*figure 5*)

Brahe enjoyed great authority, so with his thesis he favored the abandonment of the Ptolemaic system. At the same time his thesis delayed the affirmation of the Copernican system.

Tychio had learned from Copernicus the idea of the planets revolving around the Sun, but he had not found the courage to confirm the same principle also for the Earth.

However, his studies were of great help to another astronomer, Kepler. Kepler was Tycho's assistant during the Prague exile. He tried to convince Brahe to abandon the Ticonian system to adopt the heliocentric system, but without results.

At Tycho's death in 1601, Kepler replaced him in the post of Mathematician and Imperial Astronomer, in Prague.

Thomas Digges and the heliocentric model

Thomas Digges (figure 7), British astronomer and mathematician, was born in Barnham in 1546, the same year as Tycho Brahe. Thus Digges and Brahe were contemporaries. Even Shakespeare, born in 1564, lived in the same period.

Digges' mathematical background was curated by his father and one of the most famous mathematicians of the time, John Dee. Digges had the merit of being the first English supporter of the Copernicus theses. (*figure 6*).

In 1572 also Digges, like Brahe, was able to observe the "Stella nova". He published the diary of the supernova observations in 1573, in the work "*Alae sive scalae mathematica*". Digges' observations were not qualitatively inferior to those of Brahe. Indeed, he seems to have been more precise in calculating the position of the supernova.

Digges and Brahe held epistolary ties so they had the opportunity to exchange views on the respective visions of the universe. These visions were decidedly contrasting. Probably, however, what exacerbated the relationship between the two was the great recognition obtained by Brahe for his observations of the "Stella Nova". At the same time, Digges' work was practically ignored.

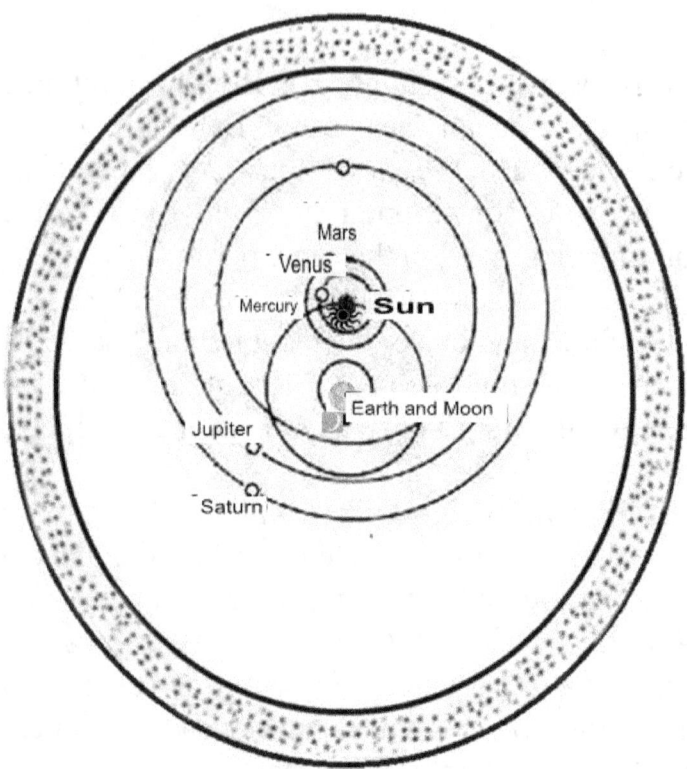

TYCHO BRAHE MODEL
The earth is at the center of the universe. The planets revolve around the sun.
The Sun rotates around the earth.

Figure 5 - The vision of the universe according to Tycho Brahe, decidedly inspired by the Ptolemaic model. The planets revolve around the Sun, but this revolves around the Earth, which remains at the center of everything.

The universe is intelligent. The soul exists.

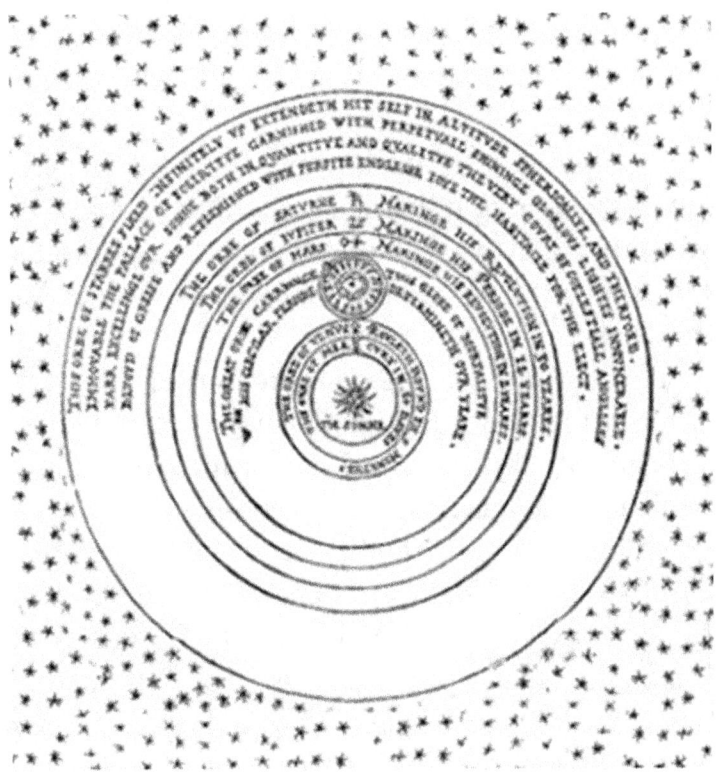

MODEL OF THOMAS DIGGES

The Sun is at the center of the Solar System. All the planets rotate around the Sun. The Earth also rotates around the Sun.

Figure 6 - Digges' vision of the universe has a decidedly Copernican orientation. The Sun is at the center of the system and all the planets revolve around it. Like the other planets, the Earth also rotates around the Sun, on the third orbit.

Therefore, when Shakespeare identified the person of the astronomer Tycho in the usurper Claudio, he did it to satisfy his personal desire for justice.

Certainly the friendship between Shalespeare and Digges came into play. According to some biographers, the two lived near one another. Probably between the two and their families there was knowledge and acquaintance.

Considering this, it is possible that Shakespeare assigned the roles of tragedy using a component of partisanship. Tycho, takes on the role of Claudio, the usurper. Digges takes on the role of Hamlet, to which his father and mother have been removed

But surely Shakespeare's choice was based on a deeper conviction. He believed that Digges' position on the reality of the universe was more correct than that of Brahe. The doubts and the problems of Hamlet represented the difficulties of Digges in supporting his thesis, considering that Brahe enjoyed greater listening and consideration.

Among the dialogues of Shakespeare's tragedy, the one between Hamlet and the two emissaries sent by Claudio is relevant. The two want to convince Hamlet that Denmark is a great place to live. But the prince declares:

"Denmark is all a prison".

Rosencrantz replies:

"Say that because you are ambitious. Denmark is too small a space for a mind like yours ".

To which Hamlet replies:

"O God, I could be bounded in a nutshell and count myself a king of infinite space".

The positions of the two astronomers clearly emerge from this dialogue. Tycho believes that the whole universe consists of the Earth at the center of stars that rotate around it. He hypothesizes a limited and finite universe in its dimensions.

Digges, on the contrary, proposes a universe without limits, where the Earth is only a planet equal to infinite others. In Digges' vision, man, although confined to an insignificant planet, can feel himself lord of something much larger, that is, of an infinite universe.

To conclude this excursus in the story of Hamlet and its author we can summarize the terms of the dilemma. Obviously this was beyond explicit narration. The dilemma that distressed Shakespeare was related to the essence of the universe. Does man live in a confined place, with precise boundaries, or in an infinite place in every direction?

Some might think that this is a dispute of the sixteenth century, that is, something that no longer matters. Well, he is very wrong. This dispute has not yet been resolved.

"De l'infinito, universo e mondi"

Man has no limits. When he realizes it, he will have gained freedom in all the worlds.
(Giordano Bruno, philosopher)

Giordano Bruno, the philosopher of the infinite

So, does the universe have a boundary or is it infinite? To deepen the controversy over the size of the universe we cannot ignore a character who did not hesitate to take part in the discussion knowing that he was putting his life at stake. I'm talking about Giordano Bruno (figure 8). Bruno was neither a scientist nor an astronomer, so he approached the subject from a completely different point of view. He was a Catholic religious of the Dominican order.

Bruno was a contemporary of all the other characters mentioned above. He was born in 1548 and died in 1600. However, it is unlikely that he had cultural relations with the characters mentioned, because he lived in different places.

His religious condition and the environments in which he published his ideas did not allow him to enjoy the freedom of thought that Copernicus and Digges had enjoyed.

Bruno clashed all his life with the ideas of Catholic theology. These ideas definitively defeated the philosopher on 17 February 1600, when he was burned at the stake in Piazza Campo dei Fiori in Rome.

The universe is intelligent. The soul exists.

Bruno's real first name was Filippo. He had received this name to honor the heir to the throne of Spain Philip II.

As regards his origins, Bruno gives this information during the interrogations to which he was subjected. I report the information with the wonderful language of 1600 Italy. It is the language that Bruno uses in the writing of his books and is preserved in the publications even today

"Io ho nome Giordano della famiglia di Bruni, della città de Nola vicina a Napoli dodeci miglia, nato ed allevato in quella città, e più precisamente nella contrada di San Giovanni del Cesco, ai piedi del monte Cicala, forse unico figlio del militare, l'alfiere Giovanni, e di Fraulissa Savolina, nell'anno 1548, per quanto ho inteso dalli miei".

(My name is Giordano and I belong to the Bruni family. I was born in 1548 in Nola, in the district of San Giovanni del Cesco, twelve miles from Naples, at the foot of Mount Cicala. Perhaps I am the only son of a militare, the "alférez" Giovanni, and of Fraulissa Savolina This is what they told me about my origins).

Bruno's philosophy focused on the idea of an infinite universe. The universe had to be infinite as a derivation of an infinite God. Therefore this God had to receive infinite love. The universe was composed of an infinite number of worlds.

In 1565 Brunoo entered the convent as a novice with the Dominican friars. At the age of 18, on June 16, 1566, he definitively entered the religious order. On that occasion he renounced the name of Filippo, as imposed by the Dominican precepts, and took the name of Giordano.

Considering his testimonies, we understand that the reason he chose to wear the Dominican habit was not the interest in religious life.

He wanted to benefit from the cultural wealth he would find in the convent. In fact, he suffered much from the cultural poverty typical of the popular environments of the time.

The first time he entered the small room of his convent he threw away all the images of the saints he found. He kept only the crucifix.

The possibility of drawing on a vast culture was not lacking in the convent of San Domenico Maggiore. Bruno knew that the convent had a very rich library. But he was very upset when he learned that the books of Erasmus of Rotterdam were forbidden. He didn't give up reading them. He obtained the banned books and studied them secretly.

Thus his education benefited authors who were often forbidden for a Dominican friar. Among others Aristotle and Thomas Aquinas, but also Marsilio Ficino, Raimondo Lullo and Nicola Cusano.

Unfortunately, his independence of thought went too far for a friar, when he came to raise doubts about the dogma of the Trinity.

He was denounced to the Provincial Superior Domenico Vita, who instituted a trial against him on the charge of heresy. Bruno left Naples and moved to Rome. In this city he abandoned the Dominican habit and resumed his original name of Filippo.

Figure 7 - Thomas Digges was the first English supporter of the Copernicus theses. The Polish astronomer, supported by Digges, placed the Earth in motion around the Sun.

The universe is intelligent. The soul exists.

Figure 8 - Contrary to the vision of the era, in his work "De l'infinito" Bruno supports the infinity of the universe and the existence of an infinite number of worlds. He was tried by the Inquisition on 17 February 1600, and was later burned at the stake in Rome in Campo dei Fiori

From that moment on, Bruno passed through countless journeys mostly in foreign countries.

His escape ended in Venice. He had imprudently gone to that city on the request of the Doge Giovanni Mocenigo. He had attracted him with a request to be educated. He claimed to want to study astronomy and the art of memorization, in which Bruno was an expert. Unfortunately Mocenigo was an agent of the Inquisition. On May 23, 1592 he had Bruno arrested and had him transferred to Roman prisons.

Giordano Bruno and his idea of infinity

Bruno expresses his idea of infinity in various works, but the most significant is "*De L'infinito, universo e mondi*" published in London in 1584.

According to the dominant theory of the era in which he lived, the universe was a place of finite dimensions, with the Earth at its center. The Sun and the other planets, instead, constituted a system of spheres rotating around the Earth. On the surface of the last sphere there were the fixed stars. These stars were unknown objects. No one knew the limits of their extension. No one knew what was beyond the fixed stars. But no one bothered to learn more about the fixed stars because this would have been of no use.

After all, even today very few people ask themselves what was before the Big Bang. All that can interest man is contained within the limits of time and space. These two dimensions did not exist before the Big Bang. We have no tools to represent a reality without space and time. Therefore, in Bruno's time, the only object worthy of interest was the Earth, especially since it represented the center of everything.

Contrary to this vision, in his work "*De L'infinito, universo e mondi*" Bruno hypothesizes a different reality, composed of an infinite number of worlds. In the "*First dialogue*" of this work Bruno maintains that the universe is infinite, because God, who generated it, is infinite.

> "Thus the excellence of God is magnified, the greatness of his empire is manifested: he is not glorified in one, but in innumerable Suns; not in an Earth, in a world, but in ducento mila, I say in infinite. ".

Previously Bruno had written the opera "La Cena de le ceneri". This work, published in London in 1584, is a philosophical dialogue on nature. In this volume Bruno refers to the Copernican theory. He proposes a universe in which the divine is omnipresent and the matter is in constant change but is eternal.

The universe that Bruno imagines is infinitely extended, composed of an infinite number of solar systems similar to the one we know.

One of the characters in the book is called Filoteo and he is the one who expresses the author's opinions. Filoteo disputes Aristotle's idea about a finite universe. Filoteo (Bruno) argues that if the universe of Aristotle is finite, it cannot exist. Another character, Fracastorio, confirms Filoteo's thesis using a Latin quote:

>FRACASTORIO. *Nullibi ergo erit mundis. Omne erit in nihilo.*
>(*So the world is nowhere. Everything is nothing, everything is zero.*)".

Bruno included in the work "*De l'infinito, universo e mondi*" three poems. Here I quote the last one. This composition is not a simple poetic exercise. The verses are a prophetic message directed to the persecutors who will put an end to his courageous life:

>"E chi mi impenna, e chi mi scalda il core?

Chi non mi fa temer fortuna o morte?
Chi le catene ruppe e quelle porte,
Onde rari son sciolti ed escon fore?
L'etadi, gli anni, i mesi, i giorni e l'ore
Figlie ed armi del tempo, e quella corte
A cui né ferro, né diamante è forte,
Assicurato m'han dal suo furore.
Quindi l'ali sicure a l'aria porgo;
Né temo intoppo di cristallo o vetro,
Ma fendo i cieli e a l'infinito m'ergo.
E mentre dal mio globo a gli altri sorgo,
E per l'eterio campo oltre penetro:
Quel ch'altri lungi vede, lascio al tergo".

"I feel safe against any dispute.
I spread my wings and can fly safely.
I dominate the skies of freedom.
I see far beyond my protesters.
Their visions are limited.
Therefore, as I fly, I show them their backs."

Ethics issues

*There is a concept that corrupts and confuses all others. I do not speak of Evil whose limited empire is ethics;
I speak of the Infinite.*
(Jorge Luis Borges, Argentine writer and poet)

According to the philosophy of Giordano Bruno, the Copernican universe in which we live and all the other infinite universes are placed in an infinite and homogeneous space " *che chiamar possiamo liberamente vacuo*", that is empty. In this the thought of Bruno coincides with that of Tito Lucrezio Caro, expressed in the poem "*De rerum natura*", written in the 1st century BC

Lucretius states that the universe is composed only of atoms (referring to the atomism of Democritus). Atoms move across the whole universe in an infinite dimension, that is, emptiness. Among other things, Lucretius states that even the soul of man is made up of atoms and that these, when the body dies, are dispersed to be reused by nature.

Others, instead, affirm that the universe is finite, both in time and in space. The Big Bang theory, currently recognized as valid, describes a universe initially enclosed in an infinitesimal point. Following a giant explosion, the space begins to expand and is still doing so. The expansion of space has a precise limit, measurable in light years since the Big Bang. This delimits not only space but also time. Once the universe was a point equal to zero, today it is a bubble extended for billions of light years. In the future, perhaps, it will widen its borders by extending even for billions of light years.

According to the Big Bang theory we could say that the universe is not infinite because it has a beginning and an end both in space and in time.

Therefore, today no one can say whether the universe is finite or infinite, but the question is not indifferent on an ethical level.

The problems of an infinite universe

Surely many readers have had the opportunity to buy products sold and promoted by organizations known as "Fair Trade".

The "Fair Trade" is a form of international trade that aims to guarantee producers and workers in developing countries a balanced economic treatment that respects their life needs.

Theoretically, if you buy a pound of coffee in a "Fair Trade" shop, you help a community located in a poor producing country. This community cultivates, collects and markets coffee independently. In this way the farmers are freed from the exploitation of companies, often multinationals. As is well known, these large companies often buy products in the third world by paying hunger prices.

If you buy a craft item, a food product or other goods, you give impetus to this initiative and, ultimately, you do "a work of good". You contribute to increasing the rate of altruism in a world often dominated by evil, selfishness and business. Therefore all of us, helping Fair Trade, believe we are increasing the goodness component of the universe.

But are we really sure?

In fact, if the universe is finite, that is, if it is contained in a limited space, good and evil are also present in the universe in finite quantities. In a finite universe, good and evil are measurable quantities. Therefore, with our generosity, we really increase the quantity of good by adding our drop of water to a sea that is vast but can be defined in its vastness.

Conversely, if the universe is infinite, it already contains an infinite amount of good. Therefore, no good deed can increase it.

In an infinite universe, when we buy a pound of coffee, countless other people are buying endless amounts of the same coffee in countless Fair Trade shops.

Our purchase does not increase the total amount of coffee that the endless producing communities can sell. Our purchase has no influence on the overall budget of poor coffee producers.

Moreover, the infinite universe would also contain an infinite amount of evil and, consequently, no bad action of ours could increase the evil of the universe. Therefore, doing good works, we would have no merit. But by doing bad deeds, what would we be guilty of? We would certainly not be guilty of increasing the "evil" of the world.

In truth we can argue that ethics considers and values individual actions in their intrinsically worthy meaning, and does not measure the null consequences they would have in an infinite universe.

It is not a great consolation. Moreover, no one will ever admit that we can kill a person, with the excuse that in an infinite universe there are nevertheless infinite copies.

If we kill a person in an infinite universe this has no relevance because that person will be killed countless times in infinite ways.

From the point of view of every religion or philosophy, but also according to common logic, it is absolutely desirable that the universe be finished. A clearly delimited universe, even if inserted in an infinite space, would be more tranquilizing for everyone.

To conclude, should the cosmos be infinite, the possibility of living in a well-circumscribed corner, as in a nut shell, would certainly be preferable.

Infinity in a finite space

Imagine a piano. The keys begin and end. You know the keys are 88, you have no doubts. The keys are not infinite. You are infinite, and through those keys you can play infinite music. The keys are 88, you are infinite.

(Alessandro Baricco, Italian writer)

A premonitory dream

A few years ago, by doing research on the web, I happened upon an English lady's blog. Unfortunately, I don't remember its name exactly. One page of the blog struck me particularly. In the text, the lady told an anecdote that had interested her elderly mother, whom we will call Margaret for convenience. Margaret had the habit of attending conferences given by more or less well-known characters, whatever the subject discussed. In the days following each conference Margaret transcribed her impressions in the journal. From this diary the daughter had drawn the episode I recount below.

In the 1960s Margaret had the opportunity to attend a conversation-conference. Other speakers included a young man who had just graduated in natural sciences and served at the Trinity Hall in Cambridge. His name was Stephen Hawking (figure 1).

In that period the topic of greatest interest in that kind of meetings was the origin of the universe. The debates focused almost always on the Big Bang. Indeed, at the time this theory was not yet accepted by all. At the end of the meeting the speakers met with the public and Margaret asked the young Stephen, who had impressed her with his ability to argue, how his passion for astronomy was born. Hawking was still young and certainly did not want to disappoint his University, which had organized the meeting Therefore, he did not give Margaret a hasty answer but he told her an episode of his life as a child. As always happened, the episode narrated by Hawking was written in Margaret's diary.

At the age of five or six, little Stephen was present at a conversation between his parents, Frank and Isobel, and a distinguished character. Stephen also listened to the conversation and was struck by a curious argument. The unknown interlocutor said, at a certain point, that if someone wanted to succeed in life he would have to reveal some unsolved mystery, for example the infinity of the universe or the correct interpretation of the Apocalypse.

This statement impressed Stephen. Although he was very young, the sacred fire of knowledge and the ambition to assert himself in life were already present in him.

He quickly discarded the Apocalypse option. Since it was a holy book, he would not have known how to get it. He knew he could not even ask for the book from his father, since the man was not interested in religious matters.

Therefore, he decided that he would discover the mystery of the infinite. From that moment he began to scrutinize the sky whenever he had the chance. The sky was a great free book, and it could be read without anyone's permission.

Many years later, one night, Stephen fell asleep meditating on the problem of infinity and had a surprising dream. He saw the universe as a gigantic wheel made of luminous fragments, which rotated slowly on itself, like a giant kaleidoscope.

At that moment, in the dream, he had the clear sensation of having understood what the universe was. The truth was before his eyes. He had revealed the secret of the infinite universe. He had only to stretch out his hand to grasp that mystery and appropriate it. Everything was very clear in the mind of young Hawking, beyond all doubt.

Unfortunately, when he awoke, he realized with great disappointment that the truth, so quickly grasped, escaped him as quickly.

Certainly, a big spinning wheel has no beginning or end, so it can be considered infinite. However the mystery is not solved. The problem always remains of knowing what lies beyond the outer limits of the wheel.

Stephen was left with the bitter feeling of having found the explanation of the infinite universe but of having lost it immediately afterwards. This awareness made him live the rest of his life with the desire to recover this truth.

We can accept some premises. The story told by Margaret is true. Margaret's daughter correctly transcribed the story in her blog. My memories allowed me to reconstruct the story correctly. These premises are true. Why shouldn't they do it? Based on these premises, we can say that the dream of young Stephen was an episode of synchronicity. The dream was a synchronicity because it anticipated a mission entrusted to a great man.

This synchronicity involved and inspired the young Stephen. He was able to anticipate the mystery he would pursue throughout his life, after having glimpsed it for a moment. Between Hawking and the infinite a relationship was established like the one that unites Narciso and Boccadoro in Hermann Hesse's novel of the same name:

> "Our task is not to approach, just as the sun and the moon, or the sea and the earth do not approach each other. We two, dear friend, we are the sun and the moon, we are the sea and the earth. Our task is not to transform ourselves into one another. Rather, our purpose is to get to know each other. We must learn to see and respect what he is in the other. We are reciprocally opposed and complementary".

The concept of the wheel makes it possible to evaluate the mystery of the infinite from an unsuspected point of view for our way of thinking. We conceive time and space according to a linear representation, as in the upper part of figure 9. Time and space have had a beginning and continue along an infinitely long line. In fact, you can add a unit to each number to increase it to infinity. In the same way, another line can be added to each line for an infinite number of times.

Instead, the circular representation of space-time, as can be seen below in the same figure, allows us to imagine a finite but at the same time infinite sequence of events. The parts of the wheel are free of any start and any termination.

The circular configuration of space-time also makes it possible to establish a connection between the dream just narrated and some philosophical interpretations of the infinite.

Surely Stephen wondered for a long time about the multitude of rotating colored objects arranged in a circular shape. Surely, at the end of his reflections, he realized that what he had dreamed of was a *mandala*.

The mandala

There is a symbolic sign present in virtually every culture and at all times: in the Sanskrit language its name is "*mandala*", a word that can be translated as a circle.

Figure 9 - Temporal space continuum. Above the linear representation, where space-time has a beginning and proceeds to infinity. Below the circular representation, where it is not possible to identify either the beginning or the end. The circular cycle is repeated eternally.

The universe is intelligent. The soul exists.

Figure 10 - Some mandalas. Generally these representations are colored.

Today, "mandala" is the universally widespread term used to refer to figures similar to those in figure 10. A mandala is also reproduced on the cover of the book. In most cases these are colored figures.

The symbol of the circle with special powers is already present in the early dawn of humanity. The first known mandala is a Wheel of the Sun dating back to the Paleolithic of southern Africa. Neolithic stone circles, such as Stonehenge, are well known.

Moreover, even the symbol of the spiral, purely mandalic, is present everywhere. In the symbolic representations of the Neolithic peoples, the spiral represented the Sun, the source of life.

Certainly in the symbolic elaboration of the mandala the fact that mandalic forms are present everywhere in nature has influenced. This is particularly true in the field of biology. The shapes of flowers, trees and many animals recall the circular shape of the mandala. Many parts of biological organisms, such as the eyes, also recall the shape of the mandala.

But the figure of the mandala marks the whole Universe, because we find it in the galaxies, in the shape and rotation of the planets, in the orbits of comets and asteroids and even in the event horizon of black holes.

A research was recently carried out at the universities of Northwestern, Harvard and Yale, subatomic particles sector.

Study teams conducted experiments to examine the shape of the electron. The conclusion was that the electron is perfectly round. In the final statement Gerald Gabrielse, who coordinated the research of the three universities, states:

> "Our research is very significant from a scientific point of view, because it confirms the standard model of particle physics, and excludes alter-native models.
>
> Any alternative model would have required different approaches to the study of matter and antimatter. "

According to Hinduism and Buddhism, in any circular form it is possible to see a mandala. In these Eastern philosophies the mandala often has protective powers and is a source of healing.

In the West, the idea of the protective circle is found in numerous folk dances, as well as in the children's circle.

Among the redskins of America the "Wheels of medicine" are widespread. These are stone circles in the center of which people who seek healing are placed.

Alce Nero was a shaman, or medicine man, in the Sioux tribe of the Oglala. In a series of interviews collected by two journalists, John G. Neihardt and Giuseppe Epes Brown, Alce Nero recounts his experience and his spirituality. In one of these interviews he states:

> "All things made by the" Power of the World "are done in a circle. The vault of the sky is round. The Earth is round like a ball. All the stars are round. The wind, at its peak, turns like a vortex. Birds make their nests in a circular shape. The Sun, which is round, rises and falls along the circle of the sky. The Moon does the same. Even the seasons form a great circular cycle in their succession. "

Jung and the Mandalas

Carl Gustav Jung (figure 11) devoted himself to the study of the mandala for twenty years. During this time he wrote four essays on the subject. In his memoirs Jung tells:

"Every morning I drew in a notebook a small circular figure, a Mandala. The traced design had to match my intimate condition. A little at a time I discovered what the Mandala really is. The Mandala represents the Self, the personality in its entirety. A harmonic Mandala represents the harmony of the person, when all is well."

Jung's interest in this symbol should be placed in the context of his theories on the collective unconscious and on archetypes.

According to Jung, man has an individual consciousness and an individual unconscious. However, beyond that, man can dialogue with the collective unconscious. The collective unconscious is a "psychic container" external to man. In the collective unconscious there are the knowledge and experiences of all humanity, memorized in the form of "archetypes". Archetypes have three main features:

- **The archetypes are universal**, that is, they are a treasure of all humanity.
- **The archetypes are impersonal**, that is, they are independent of the consciousness of individual persons.
- **Archetypes are hereditary**, that is, all can freely use them.

According to Jung, the unconscious part of man can be examined in two ways:

- **The personal unconscious** contains above all the complexes that manage the personal intimacy of psychic life.
- **The collective unconscious** contains the representations called archetypes. Often these archetypes manifest themselves through dreams. In fact, dreams are the best place where narration is free from the will and experience of those who dream. No one can decide which dream to do.

In his book "*Der Mensch und seine Symbole,*" Jung af-ferme:

> "The Mandala is the archetype of the inner order. The circular figure expresses the fact that there is a center and a periphery. The periphery tries to embrace the whole. The mandalic circle is the symbol of totality.
>
> When in the patient's psyche there is great disorder and chaos, the mandalic symbol may appear in dream, or in free fantasies or designs. The Mandala appears spontaneously. In these cases the

mandala is an archetype that compensates for disorder. The mandala can carry the order or it can show the possibility of the order ... ".

Jung's theory on the collective unconscious was challenged by many. Faced with the doubts of his colleagues, Jung supported his thesis with these arguments:

> "Man has developed consciousness slowly and laboriously. It was a process that led, after many centuries, to civilization. The beginning of civilization is identified with the invention of writing, around 4000 BC.
>
> However, the evolution of the human species is not complete, since many aspects of the functioning of the mind are still shrouded in darkness. What we call "psyche" does not correspond at all to consciousness and its contents.
>
> Whoever denies the existence of the unconscious supposes that our current knowledge of the psyche is total. This opinion is false. Even the assumption that we know all there is to know about the universe is false.

Figure 11 - Carl Jung elaborated the theory of the collective unconscious and that of synchronicity.

The universe is intelligent. The soul exists.

Figure 12 - Wolfgang Pauli, physicist, Nobel laureate 1945, worked extensively with Carl Jung. The two scientists sought a method to unify the roles of matter and psyche in the universe.

Our psyche is part of unknown nature and the enigmas of the psyche are infinite.

Therefore, it is impossible to define both the psyche and nature. We can only describe what little we understand about nature and the psyche. We can describe their functioning only on the basis of the little we know.

Consequently, there are considerable logical foundations for rejecting statements such as those according to which "the in-conscious does not exist". Beyond that, there is a lot of evidence accumulated by medical research.

The idea that a plant or an animal invent themselves makes us laugh. Yet many believe that the psyche or the mind invented themselves and created their own existence on their own. "

In reality, the mind has developed to its current phase of awareness in the same way that the acorn turns into oak. At the same node the saurians have gradually become mammals. The mind has continued to develop over a very long period of time. The mind still continues to develop. As a result, we human beings are subjected both to the action of inner forces and to the action of external stimuli.

These forces spring from a deep source, which is not constituted by consciousness. They are forces that are not controlled by consciousness.

In primitive mythology these forces were called "mana", or "spirits, demons and deities". Nowadays these forces are active as they have always been in the past. If these forces conform to our desires we regard them as positive feelings or impulses and congratulate ourselves for being well liked by fate.

If instead these forces oppose us, then we say we are persecuted by bad luck. Sometimes we say that some people want us badly. We also think that the cause of our misfortunes can be pathological. The only thing we refuse to admit is that we are at the mercy of "forces" that we cannot control ...

... There is no difference of principle between organic and psychic development. The psyche creates its own symbols, just as the plant produces the flower. Each dream constitutes a proof of this process ".

According to Jung it can happen that in dreams there are mandalic figures, especially during periods of greater psychic suffering. These figures are intended to establish an inner order. The figure of the mandala exerts a positive action, because it represents a rationalization, a "re-ordering" of tensions.

In fact, the mandala is composed of a precise center and reliable contours. The signs inside it are distributed in a geometrically ordered way.

The boundaries of the mandala enclose an area of security where the dreamer is under magical protection. Sheltered by the protective circle, almost like a new uterus, man is safe from any external attack. In these boundaries he feels secure and recovers the serenity necessary to seek his center, that is himself.

In the center of the mandala the individual finds the security he had lost, or even the security he no longer believed he possessed.

A student of Jung, Marie Louise Von Franz, highlights a further aspect that can be considered more important than the recovery of centrality. According to this scholar the mandala is a symbol of "new beginning" or restart. The figure of the mandala generates the necessary thrust to give shape to something that does not yet exist. Therefore, the mandala has a creative power, thanks to which it becomes possible to imagine new proposals and solutions.

After long studies, Jung came to the conclusion that the mandala can be considered an archetype of the collective unconscious for the following reasons

- **The frequency**, constancy and regularity with which these figures appear in the most diverse eras and civilizations.
- **The presence of a center** towards which the whole figurative system is oriented.

The delineation of the mandala generally has the shape of a circle, but it can also be similar to a polygon or a cross.

Often the delimitation consists of ornamental motifs such as, for example, flower petals.

Therefore, for Jung the mandala has the following characteristics:

- **Order and beauty**. The mandala represents order, but also the aesthetics of the universe.
- **Compensation**. The mandala compensates for the need to immerse yourself in a dimension that cures and dispels any disorder.
- **Peace**. The mandala allows you to find a serene spiritual dimension.
- **Mysticism**. The mandala has a mystical sense when it places man at its center. The man at the

center of the mandala is mystically located in the middle between heaven and earth. In this position, man yearns to merge in the synthesis of these two worlds.

- **Healing**. The mandala releases a healing force that does not depend on will or conscience.

- **Naturopathic properties**. Mandalas are nature's attempt to intervene in a healing manner on the psyche of individuals. They are the manifestation of a natural therapy of almost mystical origin.

In the same book "Man and his symbols" Jung says:

> "In recent times civilized man has acquired a powerful will power that he applies on all occasions. He has learned to do his work effectively without the help of liturgical songs or drums. He no longer needs to use hypnotism to act.
>
> He can also do without the daily prayer to invoke divine help. He can do independently what he wants, because he manages to translate his ideas into action. He can do it in complete freedom.
>
> Instead, primitive man was conditioned at all times by fears, superstitions and other invisible obstacles. These obstacles stood between the man and the action. On the contrary, the superstition

of modern man is enclosed in the motto "Wanting is power".

Yet contemporary man pays the price for a serious lack of introspection. Modern man does not see that, despite all his rationality and efficiency, he is still dominated by uncontrollable "forces".

The deities and demons have not disappeared at all: they have only changed their names. They keep the man in a state of incessant agitation. Divinities and demons manifest themselves through vague fears and psychological complications. Consequently man has an insatiable need for pills, alcohol, tobacco, food. Divinities and demons continue to impose a heavy burden of neurosis on them. "

The cosmic egg

When nothing existed yet, a goddess named "Wandering in wide spaces" danced in the void of the cosmos. In the absence of everything, the goddess had nothing to contemplate but herself. She was pleased with his movements and fell in love with his dance. His movements, at first slow then ever faster, became frantic to the point of generating the North wind, called Borea. This wind, wishing to mate with the goddess, became the serpent Ophion. In his turn Ophiion transformed the goddess into a white dove. In the guise of a dove the goddess generated the fruit of the union, the "Cosmic egg", from which all things originated.

This goddess is remembered by the Sumerians as "Divine Dove". Greek mythology remembers she with the name of Eurinome.

This story stems from the mythological studies of Robert Graves, a British poet and essayist.

We can add a gossip note. At one point the serpent Ophion boasted that he was the creator of everything. That was enough for the Divine Dove to feel offended. To take revenge, he broke all Ofion's teeth with a kick.

Mircea Eliade is a Romanian scholar born in Bucharest in 1907. He is an academic of vast culture. Mircea writes these words about the origin of the universe in his studies on the archaic oriental world:

> "The myth of the cosmogonic egg, attested in Polynesia, is common to ancient India, Indonesia, Iran, Greece, Phenicia, Latvia, Estonia, Finland, the

Pangwe people of West Africa, to Central America and the West Coast of South America ".

Since ancient times, references to the cosmic egg can be found among the Babylonians and Sumerians. From Mesopotamia, two millennia before Christ, the tradition spread in India and ancient Egypt. Later the myth of the cosmic egg also developed in China, in the European Celtic regions and in Africa.

The story of Eurinome is also told in the mythologies of the Pelasgians. In these traditions, as in many others, the cosmic egg is a reptile egg because it is laid by the serpent Ophion, which is probably the mythical Basilisk

For the Celts the cosmic egg, whose name is Glain, has a reddish color and has been placed on the primordial beach by a sea serpent.

In the Chinese Taoist religion the cosmic egg is described in the myth of Pangu. Pangu creates the world with the help of the horned serpent Qilin, the tortoise, the Phoenix and the dragon.

In Egypt the cosmic egg is laid by the Phoenix, a mythical bird-like creature. The Phoenix has a life-generating breath, from which the air god Shu is born. When it is about to die, the Phoenix builds a nest around itself. In this nest the Phoenix generates the fire that consumes it completely.

Figure 13 - Esoteric vision of the cosmic egg

However, from this combustion another egg is generated, which will be hatched by the Sun until the Phoenix is born again.

In the Bambara African tribe it is said that at the beginning of it all there was only one empty egg. This void was filled by the creative breath of the Spirit.

All things were born from here.

One of the most interesting traditions is the one narrated in the Hindu religion. Initially the cosmic egg or "Hiranyagarbha" floated in the primordial ocean, wrapped in the darkness of non-existence.

When the egg hatched, Brahmā introduced it to humanity through the "Om". This syllable allows respiratory emission, so in Hinduism it represents the original vital breath.

The upper part of the shell of the cosmic egg is made of gold, and from this part the sky is born. Instead, the lower half of the egg is made of silver, and from here the earth is born. This creation is cyclical: the Universe develops from the cosmic egg. Subsequently, the universe becomes corrupted until it reaches its end. From the end of a Universe another Universe is born, and then another. This series of cycles is the "kalpa".

These are just some of the mythological narratives on the cosmic egg.

Indeed, the egg lends itself very well to being considered the origin of all things. First of all, it has no edge. The shape of the egg is elliptical, however the ellipse has neither beginning nor end, just like the circle. This is why the egg can represent something that has always existed and lasts forever. The egg is a symbol of fertility and can be considered the primordial seed, the first embryo that emerges from the chaos to generate all that exists.

A Latin saying states "Omne vivum ex ovo", that is: "everything that lives comes from an egg".

In Egyptian religiosity the egg represented the cosmos, because it contained the four cosmic elements. The shell represented the earth, the red yolk represented the fire, the transparent albumen represented the water. Instead the fourth element, that is the air, was represented by the environment that surrounds the egg.

A wonderful graphic interpretation of the cosmic egg has found concretization in the mathematical idea of the Zero. The Zero is the primordial feminine archetype from which all numbers descend. This happens because the Zero is fertilized by the One.

Like the classic egg, zero represents "a nothing that produces something living".

In the alchemic vision the egg is an archetype that can bring every element back to its original condition of purity.

As already mentioned, the Egyptians associated the four elements (earth, fire, water and air) with four parts of the egg. Instead, the alchemists associated the three most obvious parts of the egg with three essential alchemical ingredients. The shell was associated with salt, the albumen was associated with almercium and the yolk was associated with sulfur.

According to the alchemist masters these three elements, combined in the right doses, could lead to the creation of the Philosopher's Stone, that is, to the fulfillment of the "Great Work". The Philosopher's Stone could transform the less noble metals into gold.

In the mandalas, the egg appears as the alchemical symbol of the "All".

As for our daily life, we can see that the symbolism of the primordial egg, generating life, is still contained in the Easter Egg tradition. In the current Christian tradition it symbolizes the resurrection of Jesus from the sepulcher. The sepulcher, similar to the Phoenix's nest, is the place where Christ is reborn, the origin and salvation of the whole universe.

In effect, however, the egg's gift is much older and even dates back to the Persians, among whom the tradition of exchanging simple chicken eggs at the beginning of spring was widespread. The egg represented a wish for fertility and abundance for the summer and autumn harvests. Obviously, those crops were vital to agriculture-based communities.

The cosmic egg and current physics

Based on what has been said, it seems that the cosmic egg is confined to mythological or alchemical stories.

However, in recent decades the idea of the cosmic egg has also infected the world of astrophysics. This has happened especially since science began to evaluate the theme of a primordial singularity. From this singularity, through the great explosion of the Big bang, the whole Universe would have been generated.

Indeed, the current astronomical conception configures an expanding universe. This theory derives from the observations of Edwin Hubble and, subsequently, from the general theory of relativity of Albert Einstein.

If we travel back through the expansion of the universe, turning it into a contraction, the universe becomes smaller and smaller. At some point we arrive at a dimension so tiny and dense that it cannot be described with any physical term, and therefore it is called "singularity".

Few know a generally overlooked aspect of the biography of Erwin Schrödinger, the Austrian Nobel Prize winner in 1933 for Physics. Schrödinger has developed a series of fundamental results in the field of quantum theory. He is known to the general public for his experiment called "cat paradox".

Beyond the scientific rigor of his studies, Schrödinger has been interested throughout his life in Hinduism and Vedanta philosophy.

In this context he has repeatedly expressed his philosophical position. According to Schrödinger it is possible that individual consciousness is only the manifestation of a global and unitary consciousness that pervades the universe.

It is not surprising, therefore, that Schrödinger wanted to apply the concept of cosmic egg to his studies in quantum mechanics, linking it to that of an expanding universe born of a "singularity".

The meeting between Jung and Pauli

Carl Jung studied the phenomenon of "strange coincidences" for a long time, and then attributed them with a "numinous", ie divine, character.

Precisely because of one of those strange coincidences, in January 1932 Jung received, in his Zurich studio, a visit from a character who would mark the rest of his life.

The visitor was Wolfgang Pauli (figure 12) an Austrian professor who taught Theoretical Physics at the Federal Institute of Technology in the same city. Pauli had decided to turn to Jung for psychological assistance because he had been the victim of a series of heavy adversities.

In November 1927, Berta Camilla Schütz, Pauli's mother, committed suicide. She was only 49 years old and had been a writer and feminist committed to socialism.

The following year Pauli's father remarried with Maria Rottler. Wolfgang had not accepted this second marriage, because Maria was a young woman of her own age.

Furthermore, in December 1929 Wolfgang married Käthe Margarethe Deppner, a professional dancer. In this way he thought of finding an accommodation on the level of feelings. Unfortunately, the marriage had gone wrong immediately and the two had divorced, after less than a year, in November 1930.

All these circumstances had created considerable psychological distress in the young professor. Despite his prestigious academic position, Pauli drank too much alcohol and spent his evenings in public places. Unfortunately it disturbed the other customers and was contentious, so often the managers of the premises were forced to drive him out.

The universe is intelligent. The soul exists.

Pauli turned to Jung, because the rational setting of his psyche, typical of a scientist like himself, made him understand, in moments of lucidity, the extent of the imbalance he was experiencing. Later he would have called that period "the great neurosis".

Jung describes the meeting with Pauli in his diary:

> "I had the case of a university professor, a very mono-oriented intellectual. The unconscious of this client is upset and very active. He projects himself into other men whom he sees as enemies, and feels terribly alone, because he thinks everyone is against him. "

However, Jung recognized in Pauli, from the first meeting, a formidable intellectual ability and a great scientific preparation in his profession, that is in the branch of physics.

Right from the start, Jung wished to establish a dialogue with Pauli on the level of science, rather than on a therapeutic level. For this reason the psychologist considered it correct to entrust the care of the patient to one of his most qualified collaborators, Dr. Erna Rosenbaum. This allowed him to take care of the patient by treating him as a friend, without playing the role of therapist.

He talked to Pauli in all those circumstances that might have something to do with their scientific interests.

More than 1500 of Pauli's dreams were recorded and analyzed during the therapy period. Jung used many of these dreams during his studies, not as a therapist but as a scientist.

Ultimately, the psychoanalytic path produced many benefits in Pauli. But it was above all the beginning of a collaboration based on a reciprocal exchange of experiences between two enlightened minds. In particular, the two investigated the possible links between Jung's psychoanalytic studies and quantum physics, of which Pauli was one of the founding fathers (he received the Nobel prize in 1945). Pauli was especially interested in the contents of the theory of synchronicity, which Jung was developing in those years.

In fact, the two sensed a deep connection between the behavior of elementary particles and many phenomena inherent in the theory of synchronicity. The particles studied by Pauli seemed to manifest an intelligence with psychic characteristics, ie not mechanistic. These were behaviors independent of the matter and the deterministic laws of traditional physics.

From 1932 to 1957 Jung and Pauli kept contacts constantly. In 1940, following the outbreak of the Second World War, Pauli emigrated to the United States, where he became professor of Theoretical Physics at Princeton. From that moment on, contacts continued through the exchange of letters.

The psychic diagram of Pauli and Jung

Between June and December 1950 Jung and Pauli, in their correspondence, completed the elaboration of a "quaternary". The goal was to represent a cosmic reality in which matter and psyche were reconciled in a common collaborative design.

It is interesting to reconstruct the evolution of the thought of the two scientists. The elaboration of the "quaternary" took place through proposals and subsequent adjustments.

All the following quotations are taken from the collection of the letters of Jung and Pauli. The complete collection is published in the volume "*Jung e Pauli. Il carteggio originale: l'incontro tra psiche e materia*" was edited by Eva Pattis Zoja and Carla Stroppa for the Moretti e Vitali publishing house.

In a letter sent by Kusnacht on June 20, 1950, Jung casually communicates to Pauli some insights into dreams he was studying. At the end of his elaborations Jung draws a schematic representation referring to one of the dreams: (See *D-1*)

Probably Jung traced this pattern only to comment on a dream. He did not imagine that Pauli would have seized the opportunity using the same scheme, which he defined as "quaternary", according to a much broader concept.

Indeed, at that time Pauli was looking for the answer to a question he could not solve. Pauli asked for Jung's opinion in a letter sent by Zollikon-Zurich on November 24, 1950:

> "... This brings me to the question whose discussion forms a major part of this letter.
> How do the facts that constitute modern quantum physics relate to the phe-
>
>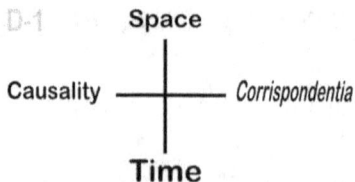
>
> nomena linked to the new principle of synchronicity? First of all, it is certain that both types of phenomena go beyond

the limits of the "classic" determinism ... For me this question is of particular importance. Since my studies are about physics, I have been discussing and reflecting on this for a year now. "

The question is addressed to Jung, but Pauli probably already had an answer in mind. In fact, after a few lines Pauli takes up the theme of the quaternary:

"To emphasize the difference between microphysics and other cases in which the psychic is involved, in 1948 I published an essay on the Hintergrundpsyche. In the article I proposed a quaternary scheme in which the two cases must correspond to different pairs of opposites. The first pair of opposites: (*see D-2*) belongs to physics:

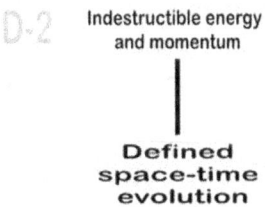

Instead, another couple: (*See D-3*) belongs to psychology:

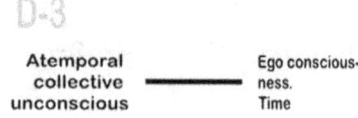

Of course I cannot say that this quaternity is suitable for the synchronicity theory.

However, my scheme has the advantage that space and time are not put in opposition to each other.

The opposition between time and space is a solution that particularly repulses any modern physicist.

Therefore, in my role as physicist, the contrast between space and time expressed in his scheme seems to me to be hardly acceptable.

In the first place, space and time do not form a true pair of opposites, since space and time can be applied simultaneously to phenomena.

Secondly, I remember that you yourself produced, in other cases, formulations in favor of the essential identity between space and time ".

Pauli does not hesitate to submit the thought of his doctor and psychologist to the most rigorous scientific principles and methodologies. However, Pauli wishes to continue the discussion in a constructive way, so he formulates a compromise proposal:

> "... So I would suggest the following compromise proposal for a quaternary scheme. The scheme I propose avoids the overlap of time and space. I believe that this solution can combine the advantages of our two schemes: (*See D-4*)

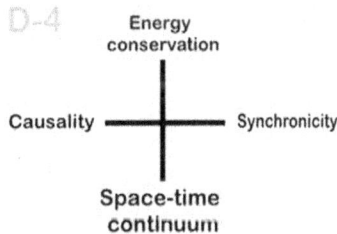

Jung replied with a letter sent by Bolligen on November 30, 1950. He formulated several arguments on the separateness of time and space, from the psychological point of view:

> "... Space and time are intuitive concepts, therefore, in an image of the intuitive world, they are eternally separated and contrary. In my scheme I take psychological criteria into consideration. In these cases these are intuitive-perceptive and not abstract concepts ".

But Jung concludes this way:

> "His compromise proposal is very welcome, because he makes the brilliant attempt to transcend intuition.
> His proposal completes the intuitive vision of the world through what is deep inside. "

However, Jung suggests modifying "randomness" and "synchronicity". . Jung thus motivates his proposal:

> "My scheme seems to satisfactorily formulate the intuitive world of consciousness. This scheme satisfies on one hand the postulates of modern physics and on the other those of the psychology of the unconscious ". (*see D-5*)

The universe is intelligent. The soul exists.

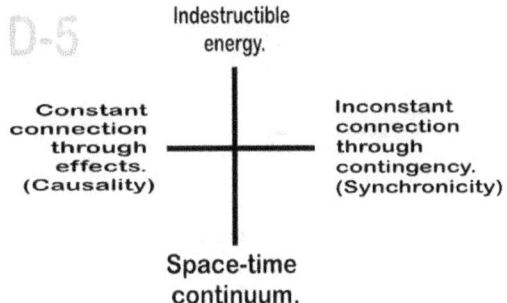

Pauli's answer came once again from Zollikon (Zurich) on 12 December of the same year. Pauli concluded the letter in this way:

> "I have no doubt. The new formulation of the "quaternary of the image of the world" that sent me is really the most appropriate expression. Moreover, this formulation corresponds almost entirely to my previous desires."

The final version of what is generally called "Pauli-Jung's psychophysical diagram" assumes the simplified aspect that I reproduce below. (*see D-6*). Today the diagram is quoted almost everywhere in psychology studies.

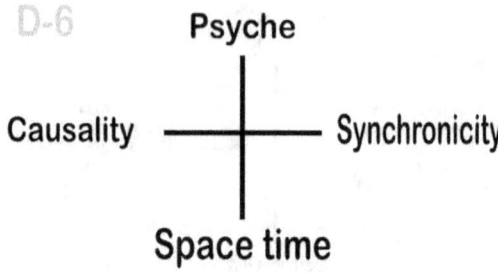

In the left-right arm of the diagram causality, or determinism, balances with synchronicity. In this way the authors hope for a collaboration between mechanistic physics and the principle of synchronicity. According to mechanistic physics (causality) every event is connected to the cause that produces it. Instead, the synchronism refers to events that are absolutely disconnected from each other. But these events (coincidences) become coherent when the protagonist gives them meaning.

The horizontal plane of the diagram balances these two opposite visions.

The vertical arm represents, above, the world of the psyche, which is balanced with the world of classical physics, placed at the bottom.

The world of the psyche is that of non-locality, where time and space do not exist. Instead, the world of classical physics is dominated by the four known dimensions, three spatial and one temporal.

Let's talk more about Mandala

Many readers will not have missed the fact that Pauli-Jung's psycho-physical diagram can be compared to a mandala. In fact, the four symmetrical arms can be enclosed in a square or round border.

Precisely for this reason, in figure 18 I present a mandalic representation of the psycho-physical diagram of Pauli - Jung. Since everything would not make sense if it were not applied to man, at the center of the mandala I inserted the icon of the Man of Vitruvius.

The central icon does not want to represent only the inhabitant of the planet Earth, but every form of intelligence present in the universe. In fact, Pauli-Jung's mandala is a universal symbol.

The infinite in the finite

Let us still consider the great genius of literature that was Shakespeare.

Although he was already popular in life, he became immensely famous after his death. His works were magnified by many influential personalities and became objects of study as well as representation. Currently Shakespeare is considered one of the most important writers in English, and the most eminent playwright of Western culture.

Thanks to his celebrity Shakespeare received many tributes also in the expressive arts. In figure 14 we can see some examples of his image and his thought reproduced in various artistic forms.

Figure 9 - Shakespeare's appearance and his works have been immortalized on paper, in marble and in the various representative forms of thought.

At the top left you can see a portrait by Martin Droeshout, an English engraver who became famous precisely for having created this work.

The portrait of Droeshout became the decorative illustration of the frontispiece of the "First Folio", the first collection of Shakespeare's complete works and was published in 1623.

Of course you see the portrait of Droeshout reproduced on paper, that is on the page of this book.

Actually, you might see the image reproduced on many other types of media, for example on fabric or ceramic.

But let's consider the example of paper. A white sheet has been printed with any technique at your pleasure. So you can see the image reproduced on the sheet. Before printing, the surface of the sheet was white. Now, after printing, the surface of the sheet is occupied by the image of Shakespeare.

Should we believe that this was the only printable image on the sheet? Absolutely not: any other image could have been printed on the same sheet. To be precise, an infinite number of images could have been imprinted on the same sheet. For example, all the famous paintings of all ages, Botticelli's Spring, Picasso's Guernica or Andy Warhol's Marylin; all the hearts carved on trees by lovers, all the uncertain drawings of children on school desks, all the scribbles that millions of men trace as they answer the phone, all the naive paintings painted by the Non-ja monkey, or by other animals, and all the surreal geometries painted by nature, like the clouds in the sky, the winding course of the rivers, the yellowing leaves or the footsteps of the crabs that run along the sand chasing the wave that brought them to the beach.

All this, and infinitely more, could find a place on that sheet. Therefore, the simple white sheet is a "matrix" on which we can imprint the entire universe.

Unfortunately it is a one-dimensional matrix. Following the multiplication of overlaps, the matrix would at first become confused, then incomprehensible and finally completely black.

In the same figure 14, top right, we can see a statue of Shakespeare sculpted by Giovanni Fontana, an artist born in Carrara, Italy, in 1820. In 1874 the statue of Fontana was placed in the center of the Leicester Square Gardens in London . The sketch of this marble statue was taken from a monument dedicated to Shakespeare by the Flemish sculptor Peter Scheemakers. This original monument was erected in 1740 and is located in the corner of the poets at Westminster Abbey.

Let's go back to the statue of Fontana, reproduced in figure 14. Marble is a three-dimensional matrix. If we imagine a single block of marble, both the statue of Giovanni Fontana and that of Peter Scheemakers could have originated from this block.

Indeed, from the hypothetical marble block that we are imagining, all the statues in the world could have originated, from Michelangelo's Pietà to any modern art sculpture. Each marble block can contain all the Paleolithic Venuses found in various European locations, such as the Lady of Brassempouy in France, or the unnamed ones found in the excavations of the Balzi Rossi in Ventimiglia. The same block of marble can contain the statues of all the goddesses and divinities of every time, the capitals of each column, the floors of each building, the Victory of Nike, Amore and Psyche by Canova or The Kiss by Auguste Rodin, the Biancone of Piazza della Signoria in Florence or the veiled Christ of the Sansevero Chapel in Naples.

In fact, each block of marble is a matrix that potentially contains all the statues of the universe. As long as the block remains intact, it contains an infinite number of statues, each mystically outlined in the structure of its matter and potentially ready to emerge.

In front of each white sheet or in front of each block of marble the artist sees and imagines his project, choosing it among infinite possible projects, and realizes it. Unfortunately the artist's work damages the matrix and excludes the possibility of creating a different work.

However, when an artist begins to engrave the stone, every marble flake that separates from the block continues to contain potentially infinite statues. Similarly, if a sheet of paper is fragmented, each fragment can contain infinite images.

Below left, in figure 14, we can see the frontispiece of the volume "Hamlet" which is probably Shakespeare's most famous work.

The drawing depicts a book, but the work does not come from the book. The matrix of the work is another. The book is only a useful vehicle to represent the fruit of a creation of the author's imagination. Therefore, in the case of the tragedy of Hamlet, the matrix is not the book, but the mind of Shakespeare, his thought.

Thought is a matrix in which infinite works can exist without one hindering the other. While Shakespeare meditated on the realization of Hamlet at the same time in his thought there existed all the works written by him, and naturally also those not written, those just sketched or those just imagined in the glare of a single instant or in the fragile fragment of a confused dream.

Shakespeare's mind could contain infinite thoughts and his thought could create endless stories. No story was erased when another emerged. All the stories remained alive and present, awaiting their moment.

The thought of Shakespeare, but also the thought of every man, is not two-dimensional or three-dimensional, but has infinite dimensions. Thanks to this, thought can contain infinite suggestions, ideas and stories, without any of them damaging the others.

In figure 14, lower right, the sculptor highlighted a phrase by Shakespeare taken from the Twelfth Night Act IV, Scene II. The inscription, which needs no comment, is "There is no darkness but ignorance".

In this case, only an extremely small part of Shakespeare's thought is highlighted. However, this short sentence has no less dignity than other works considered in their entirety. . This confirms that thought can create and contain at the same time infinite works, or even tiny fragments of the same works. These are fragments that can emerge and manifest only for a brief moment. After these brief appearances the ideas remain hidden, they can be forgotten but they never die.

The thought of a man is a bounded, absolutely personal matrix, confined in the mind of a single person. However, thought is a matrix that can actually contain all infinity at the same time.

Thought is a "finite" that contains the infinite

If we want to imagine a "finite" universe we cannot ignore the concept of infinity, also because we do not know how to give logical answers to the problems that arise.

Establishing that a thing is over means being able to set a boundary, beyond which the thing no longer exists.

Well, however, beyond this border what is there? Some might say "emptiness", "nothing". These are not satisfactory answers. In fact, even emptiness and nothingness are "something", they exist as such. They exist because we ourselves invoke their existence. Therefore, beyond the first "nothing" we should imagine a second "nothing", and then a third, to infinity.

If we want to take a practical demonstration as an example, it is sufficient to refer to the highest number that we can imagine. If the universe were finite, it would end up at that number. And yet everyone will be able to add 1 to that number, moving the end of the universe a little further. But then it will still be possible to add another unit, and then another, practically ... to infinity.

Continuing to add unity to our number, we show that the universe has no boundaries, so it is infinite.

On the other hand, we can imagine a person's thinking. This thought is definitely over, because its boundaries lie in the mind of the same person. Of course, by living, this person continues to add notions, knowledge and insights to his thought. Do you believe that someone, at some point, could say to him: "Enough, you cannot add other notions, his thought is full."?

No, in any case there would still be room in his mind to add the Bushman alphabet, an essay on the early flowering of violets, the image of a northern lights, the sound of a coconut falling to the ground, and so on to infinity.

Even our thought is like Shakespeare's. Although it is "finished" it can contain the infinite. Probably our physical mind, that is in our brain, can be considered similar to a sheet of paper or a block of marble.

Theoretically our physical brain can physically move some ideas to remote areas, so we say it can "forget". But our mind does not forget. Our mind will always be able to give life to every thought element, even if it has been forgotten or removed. Our mind will always be able to wander between infinite ideas, even those that are not thought.

It is not necessary to reason or elaborate formulas to prove this truth. It's very simple. "You know that".

Cosmic Thought

If we want to connect the considerations just made to the universe in which we live, we can begin by imagining a "finite" object, for example a nutshell. We place this object in the frame of an infinite Cosmic Thought.

Figure 15 - The cosmic thought represented in a metaphysical vision. It is an infinite place in which infinite universes are generated. Each universe is finite, and is independent of the others. In the same way, an infinite walnut tree generates an infinite number of nuts each independent of the other.

Just as Shakespeare's thought was able to create infinite works, each independent and each really existing, Cosmic Thought can generate an infinite number of universes. Imagine an infinite walnut tree. This tree can produce an unfinished number of nut shells. Each shell occupies its finite space without being aware of the existence of others.

Figure 15 depicts the Cosmic Thought in the form of a walnut tree that generates endless nuts. Each nut contains a distinct universe. From each nut a new tree can be born, that is, a new infinite Cosmic Thought that generates infinite universes.

This is not important for us since, even if it were true, we would never have the proof and we would never suffer any consequences. Even in the case of time and space travel, the places that can be reached would still be confined to our Universe.

The multiverse theory

"One of the fathers of quantum physics, Hugh Everett, argued that a subatomic particle can move simultaneously both to the right and to the left. The observer selects one of the two possibilities. However, the other continues to exist elsewhere, ie it survives in one of the parallel universes.

(Sonia Fernández-Vidal, CERN and Los Alamos)

Probably, beset by daily commitments, we are not very passionate about new scientific theories, but there is one that is truly surprising: in addition to our universe, there could be many other so-called "parallel" universes.

So far this possibility has emerged only in science fiction books and films, but we know that often these preceded scientific knowledge. In fact, it is not impossible that parallel universes really exist and are not merely the result of the imagination of film writers and screenwriters.

The hypothesis states that outside of our space-time there may be parallel dimensions, which would give life to more universes.

The most credible theory about the birth of parallel universes holds that successive waves of cosmic inflation have occurred since the Big Bang. That is, the universe would have undergone many cascading processes of dizzying expansion.

These inflationary events would have given rise to a bubble super-universe, a fractal structure in which each bubble would represent a cosmic unity of its own. Therefore, reality would be composed of many alternative universes to ours. These universes would not be very different from each other, and of course each of them would not be infinite.

The set of these universes, being the sum of "finite" elements, would in turn be "finite". However, it could contain infinite expressions of reality.

This hypothesis has been formulated for some time, and is known as The multiverse theory. Propelled by several physicists, it has recently been supported with authority by Stephen Hawking.

Shortly before his death on March 4, 2018, Hawking proposed an interesting interpretation of the multiverse. He did it in a scientific essay entitled "*A Smooth Exit from Eternal Inflation?* ". The work is the result of a long collaboration between Hawking and the Belgian physicist Thomas Hertog.

A recent study, conducted in collaboration between the University of Durham (Great Britain) and the University of Sydney (Australia) has added astonishing specifications to the theory of the multiverse. According to this research, parallel universes could even host life. The detailed results of this study have been published in the MNRAS (*Monthly Notices of the Royal Astronomical Society*), one of the most important scientific publications in the field of astronomy and astrophysics.

The theory of the multiverse

The universe is intelligent. The soul exists.

According to some scientists, at the base of the birth of several universes there would be the "dark energy", a mysterious and completely hypothetical force in that, with current means, one cannot prove its existence. However, this form of energy, if there were one, could solve many problems. For example, its homogeneous diffusion in space could generate negative pressure capable of justifying the accelerated expansion of the universe

However, the equations that calculate the total dark energy generated by the Big Bang give a result much higher than that required for acceleration.

So what is the rest of the dark energy for?

There are two possible answers:

- Our universe accelerates far more than we believe.
- The excess dark energy is used for other purposes that we do not know.

The multiverse theory was born precisely to justify this inexplicable overabundance of dark energy. The excess quantity would be distributed in other universes, parallel to ours.

However, upstream of this, we must make a very important consideration. The universe in which we live is very fortunate. The luck lies in the fact that our universe contains the right amount of dark energy necessary to produce exactly the acceleration to which we are subjected. The amount of acceleration is present in the precise quantity necessary to allow the evolution of life. It is an absolutely precise value, which allows us to: win the lottery of life every day.

It is evident that with different acceleration values the whole evolution of the cosmos would have been different, including the conditions for the development of life on the planets.

For example, a greater or lesser acceleration in quantities of only one per thousand would have given shape to an alien universe. The periods of rotation of the planets, the forces of gravity, the distribution of galaxies and solar systems would have been different.

It is almost certain that on Earth we would not have had water and atmosphere. Our planet could have been a gaseous or rocky desert, like almost all the other planets we know. Perhaps the Earth would not even exist. Precisely that acceleration, precise to the thousandth, has meant that life could develop on Earth.

Quantum physics is the mother of the multiverse

The true origin of the Multiverse Theory must be sought in the development of quantum physics.

An American scientist, Hugh Everett, proposed in 1957 the so-called "*Many-worlds interpretation*". Everett developed the theory in the field of studies and experiments on quantum behavior of matter.

According to these studies, every time the world faces a choice at the quantum level, the universe divides itself into two.

To fully understand this theory it is necessary to take a step back and examine some basic rules of quantum physics, in particular the "superimposition principle".

This is one of the basic principles of quantum physics, probably the most important. This principle states that:

> "Any two (or more) quantum states can be added together and the result will be another valid quantum state".

Probably this definition seems rather obscure for non-experts, so we can explain it with an example, which we take from traditional physics.

Let's imagine throwing the classic stone into the pond. Around the falling point of the stone there is a series of concentric circles that slowly widen on the surface of the water.

We now throw a second stone near the first: this stone also produces a circle of waves that expand.

At a certain point, the circle produced by the first stone meets the one produced by the second stone, and the two circles merge. More precisely, the two circles add up.

The first and second circles continue to exist simultaneously, but in addition a "interference figure" is formed, given by the sum of the two circles.

These phenomena occur whenever we are dealing with waves of any kind, not just waves on the surface of the water.

Interference figures also develop in the case of acoustic waves, when several distinct noises are added together. It doesn't matter if the sum produces the harmony of an orchestra or the din of the crowd of screaming people in the exchange stock exchange room.

Also in the radio frequency range two or more waves can add up: Each radio or TV set is connected to the antenna. The antenna simultaneously picks up all the frequencies present in the atmosphere. The result is an incomprehensible confusion. Therefore, the frequency mass captured by the antenna must be processed by the decoder, which sends only the frequency selected by the listener to the loudspeaker.

Consider another hypothesis. If we throw pebbles at a wall, the wave phenomenon is not generated. The stones that impact against the wall do not generate figures of interference, but produce small lesions in the wall itself. Each lesion is distinct from the others. Among the impacts, interference phenomena are not generated.

This just helps us to understand an experiment, called the "double slit" experiment, performed for the first time in 1801 by the British Thomas Young. Young's intent was to understand if light is made of particles or waves.

Young threw beams of light at a barrier. The barrier had two holes or slits and was placed in front of a sensitive screen.

Young's idea was this:

- If the light is composed of particles, each particle passes through one or the other slit. The particle continues beyond the slit, reaches the sensitive screen and leaves a sharp image similar to the impact of a bullet.

- If the light is a wave, it expands until the circles reach both slits and pass through them. The wave continues to expand even beyond the cracks. Beyond the two slits two wave systems are generated. The circles of these systems expand and overlap. As a result, interference patterns are formed on the screen.

The experiment confirmed the second hypothesis. Interference figures appeared on the sensitive screen.

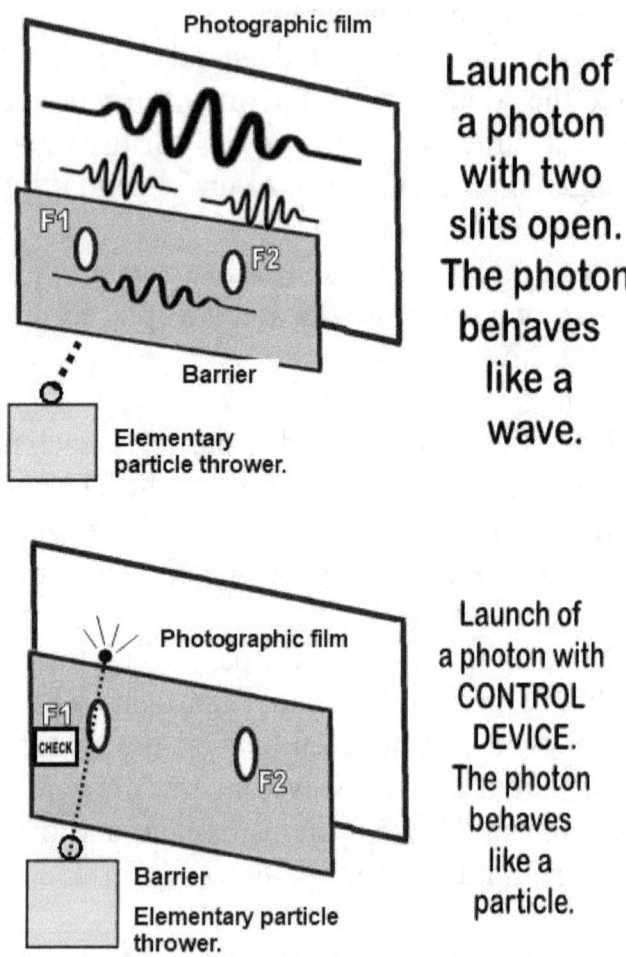

Figure 16 - The double slit experiment

Therefore, Young has concluded that light is a wave.

In recent times, many research laboratories have decided to repeat the experiment. Technical progress has made it possible to improve procedures. Instead of throwing light beams at the barrier, the scientists launched the single photons, that is, the elementary units of light. The aim was the same, that is, to understand if the photons are particles or waves.

The experiment, repeated countless times, always produces the same result which is decidedly surprising. Let's examine it in detail.

The apparatus is similar to that of the original experiment. Includes a barrier with one or two slits and a rear screen composed of a light-sensitive photographic film.

First phase. A slit

Single photons are thrown against a barrier with a single slit. The photons pass through the slit and produce single points on the back screen, similar to the impact of a bullet.

This suggests that photons are particles: in fact, beyond the barrier, they do not produce interference figures as a stone that hits water would have done. Instead, after passing through the only slit, the photons proceed straight to the photographic screen and leave the sign of an impact.

Second phase. Two slits

The surprise comes when the experimenters throw a single photon against a barrier with two slits (figure 16, top). According to expectations, that single photon must pass through one or the other slit. Afterwards, he must leave the mark of a bullet behind the crack he has passed through.

In fact, however, something incredible happens. That single photon crossed ALL THE TWO slits, and it is proved by the fact that interference figures are created beyond the barrier.

This experiment puts us in front of one of the deepest mysteries of quantum physics.

Third phase. The role of the observer

But the surprises are not over. To better understand what happens, we decide to place a sensor behind the F1 slit. If the photon passes through that slit, the sensor records the passage. In this way we know that the photon has passed through the F1 slit. (figure 16, bottom). We still throw photons, but at this point an even more extraordinary event occurs. The photon seems capable of "knowing" that we are controlling it and reacting in different ways:

Situation 1 - Two slits, one with a control sensor:
The photon passes through only one slit, the one equipped with a sensor. The sensor signals the passage of the photon. The photon produces the impact of a projectile on the sensitive screen. No interference figures are created.

Situation 2. Two slits without the control sensor.
The photon passes through both slits and produces interference patterns on the screen.

The result is unequivocal: a single photon can behave both as a particle, producing on the screen the sign of a projectile impact, and as a wave, producing interference figures.

The behavior is determined by the presence of a control sensor. This means that the observer, that is the person who organizes the experiment, can determine the behavior of the photon.

A single photon that passes through both slits represents an incomprehensible and unreal situation.

However, everything returns to normal if that photon is observed. In the presence of a control device, the photon passes through a single slit.

Let's try to express the concept in other words.

a) When the photon is not observed it is simultaneously present in all possible states:
- State 1: The photon passes through the slit F1.
- State 2: The photon passes through the slit F2.

b) When the photon is observed, it passes through a single slit.

As we say technically, the photon observed "collapses" into one of the possible states:
- Single state: The photon passes through the F1 slit or through the F2 slit.

This applies to all subatomic particles, not just photons.

Furthermore, we must consider that if we open ten slits on the barrier, the unobserved particle passes through all ten. Conversely, the observed particle "collapses" and passes through only one slit.

This experiment is incredible and important, to the point that we can still dwell on the results.

Imagine a barrier with three slits, F1, F2 and F3.

Starting situation:
- At the moment it is launched, the particle does not know that a barrier has been placed in front of it.
- The particle does not know if there are any cracks in the barrier, and how many there are.

- The particle does not know whether, in the presence of barriers or cracks, its passage will be observed or not.

However, everything that happens along the way suggests that the particle knew it, which is inexplicable.

In fact the particle behaves in the beginning of the path as if it knew what it will encounter in the path.

Carrying out the experiment WITHOUT the control sensor.

Arrived on the barrier, the particle crosses all three slits, therefore it finds itself simultaneously in the F1 slit, in the F2 slit and in the F3 slit.

It is said that the particle is in "superposition of states". All three states in which it is found are simultaneously true. We cannot say that there are three particles. There is only one particle, but all three states exist at the same time. It is an incredible ubiquity.

Carrying out the experiment WITH the control sensor.

Previously the experimenter set up a sensor at the exit of the F2 slit. Quantum collapse occurs, ie the particle only passes through the F2 slit.

The other two states "disappear". We can say that they collapse on the state under observation.

There are two questions. The first question is this: how he knew the particle, before he met the barrier, which state should he assume?

The second question is this: the particle was potentially present in three states, but collapses into the F2 state. What happens to the other two states, F1 and F3? Does it really happen that the states F1 and F3, at the moment they collapse on F2, cancel and cease to exist?

In the answer to this question lies the whole knot of the existence of the multiverse.

Most scientists say that F1 and F3 were just probabilities, so they cease to exist as soon as the alternative probability F2 becomes real.

Yet another scientist, Hugh Everett, supports a different theory. According to him, all three probabilities really exist. When the observer causes the collapse of a probability in our reality, that is in our universe, the other two collapse into other universes.

Therefore, according to Everett, every state of probability present in our universe involves the existence of other universes. Of course, even our universe could be the condensation of a probability born elsewhere in the cosmos. Our universe could simply be one among countless other probable universes that have become possible.

Actually our universe can be a tiny nutshell in a cosmic reality that could not be more infinite.

As fanciful, the concept of the multiverse has not been confined to Everett's assumptions. In recent decades the concept has also been affirmed in other scientific theories, especially the "string theory" and the "chaotic inflation", or "bubble theory".

The hypothesis of the multiverse, today, is a source of disagreement in the community of physicists, who consider it too risky and place it among the border sciences.

However, the supporters are numerous and qualified. Among them is Stephen Hawking, whom we have already known on these pages and about whom nothing needs to be added. Supporters include Steven Weinberg, 1979 Nobel Prize-winning physicist, Brian Greene, a professor at Columbia University and one of the most important scholars of string theory, Neil Turok, a South African expert on string theory, Max Tegmark, a Swedish cosmologist professor at the Massachusetts Institute of Technology, Alex Vilenkin, Russian, author of research on cosmology, cosmic inflation, dark energy and quantum cosmology.

We must also remember Andrej Linde, Professor of Physics at Stanford University, known to be the father of the theory of chaotic inflation. In addition to these there are many other scientists that should be considered reliable, because they have made important contributions to current scientific knowledge.

How many types of multiverse exist?

In 2011 a book by Brian Greene entitled "The Hidden Reality: Parallel Universes and the Deep Laws of the Cosmos" was published. The author lists nine types of plausible parallel universes, that is, such that they are not the result of fantasies or ideas that are far-fetched but compatible with scientific theories. These theories are not confirmed, but are born in accredited environments that make them worthy of consideration.

Brian Greene is an American physicist, one of the most famous supporters of string theory. Among his hypotheses on the multiverses I report here only some. I chose those that may be interesting for the topics covered in this book. Those who want to learn more about the subject can easily find Greene's book for sale.

The multiverse landscape

The "landscape" multiverse consists of "Calabi-Yau spaces". These spaces are landscapes on which the multiverse appears. These spaces are related to string theory, which predicts the existence of a number of dimensions from 10 to 26, that is many more than the four we know (length, width, height and time). The other dimensions would be hidden and "rolled up" in every point of space-time. However, even if they are hidden, this dimensions can change their energy levels due to quantum fluctuations. In this way new spaces are created, each with different laws.

Currently the expansion of the universe is accelerating. This acceleration would be due to the "cosmological constant" that is to a dark energy that permeates the space. Currently we do not know what dark energy is, and we do not even know why it has its specific value. Some attribute this value to the anthropic principle, that is, to the idea that the universe was designed to allow the existence of life and man.

Applying the anthropic principle we understand that ours is only one of the many possible universes. The other universes can have equal or different dark energy values. These values are custom designed for the life form that must be developed in that universe.

The quantum multiverse

In quantum mechanics, matter is composed of particles or waves. This means that it is not possible to determine the velocity of a particle and its location at the same time.

In other words, we cannot accurately know the spatial location of a particle. Therefore the particles are described through the Schrödinger equation. This equation determines the probability that a particle can be found in one place rather than another. Only when a particle is observed does it "collapse" and its position becomes certain.

In the quantum multiverse a new universe is created whenever an event has different probabilities.

This is Hugh Everett's "Many Worlds Interpretation". This interpretation predicts that every measurement or every observation causes the division of our reality into many worlds.

The particle, being a wave function, is subjected to the quantum principle of "superposition of states", so it can be simultaneously in two or more different places. While in our universe the particle is at a certain point, in other universes, temporarily, it can be found at different points.

The simulated multiverse

This type of multiverse exists in complex computer systems that simulate the functioning of many universes. According to this interpretation we live in an artificial universe created as a simulation on a super-advanced computer.

Probably, in the distant future the advances in technology will allow the creation of computers capable of simulating an entire universe. However, it is not clear whether a being like man, who is endowed with consciousness, can be created and simulated by a computer.

The physicist and mathematician Roger Penrose has demonstrated, based on Gödel's incompleteness theorem, that some functions performed by our brain are impossible to reproduce for any computer. Consequently, at present this multiverse hypothesis is completely excluded.

However, if in the future it will be possible to create simulated universes, in every simulated universe there will be technological civilizations that can, in turn, create simulated universes.

The ultimate multiverse

This multiverse would contain every mathematically possible universe, each with different laws of physics. This is the most philosophical category.

The final multiverse derives from the "Principle of fecundity" theorized by the American philosopher Robert Nozick. According to this principle every possible universe is real. In truth, this principle has much older origins and goes back to Plato who called it "Principle of fullness".

The brane multiverse

This multiverse derives from the M-theory "*The Mother of All Theories*". Hawking worked on this project in the last years of his life. The theory M attempts to unify all existing theories and all the fundamental interactions of matter (gravity, nuclear forces and electromagnetic force). According to the M theory, every universe would be a three-dimensional brane. Let's take an example. If the three-dimensional branes are slices of bread, then the multiverse is the loaf that includes all the slices.

Aesthetics of science

These multiverse models cannot be verified experimentally and this places them, for now, in the ambit of philosophy or metaphysics.
However, experience reminds us that many scientific discoveries have arisen from eccentric or extravagant intuitions. Of course, this happens only in a few special cases. Indeed, this rarely happens

So why did scientists engage in the development of certain theories, knowing that it will be difficult to find confirmation?

Probably, the scientist is like an artist. He feels the need to express, with the tools he possesses, the aesthetics of his thought.

In the same way that the aesthetics of art exists, there is also a real aesthetic of science. This feature completes the mathematical beauty and logic of scientific studies.

The Indian physicist Subrahmanyan Chandrasekar, Nobel Prize in Physics 1983, wrote the essay "*Truth and Beauty: Aesthetics and Motivations in Science*". In the introductory comment we can read this statement:

> "A great scientific theory is also a work of art. For the greatest scientists, beauty has always been one of the objectives to be achieved, a guide on the path to truth ".

John Sullivan, author of the biographies of Newton and Beethoven, writes in "*Athenaeum*" in May 1919:

> "The primary objective of scientific theory is to express the harmonies that

are observed in nature. As a result, these theories must have an aesthetic value. Indeed, the measure of the success of a scientific theory is the measure of its aesthetic value.

Indeed, a scientific theory is all the more valid, the more it introduces the harmony where there was chaos.

The justification of a scientific theory and of the scientific method can be found in their aesthetic value. The reasons that guide the scientist are, from the beginning, manifestations of the aesthetic impulse.

Science can be considered inferior to art only when it is an incomplete science.
"

Intelligence at the center of the universe

*It is not the Matter that generates Thought,
but it is the Thought that generates the Matter.*

(Giordano Bruno, philosopher)

The role of the observer

Now we have to go back to the double slit experiment, described above, to make some important observations about the role of the observer. We are talking about the figure of the observer understood according to quantum theory.

The most important consequence of the double slit experiment is that the observer can determine the behavior of the photon simply by observing it. In symbolic terms we say that the "look" of the observer modifies the behavior of the matter.

This happens not only with photons, but with all elementary particles, such as protons and electrons.

John Wheeler was an American physicist. He was a charismatic figure in the physics of the 30s and 40s. Many famous physicists have grown up under Wheeler's leadership, including Richard Feynman.

Among other things, Wheeler coined the term "*wormhole*" to indicate the space tunnels that allow the different regions of space-time to be connected.

Wheeler believes that the involvement of the observer in the subatomic reality is certainly the most important aspect of quantum physics. Therefore, Wheeler proposes to replace the term "observer" with that of "participant". He expresses this belief in a famous quote:

> "The measurement changes the state of the electron. After the measurement, the universe is no longer the same. To describe what happened, we need to eliminate the old word "observer" and replace it with the new term "participant". In a way, the universe is a universe based on participation ".

In the level of elementary particles the consciousness of the observer can participate in the functioning of matter, indeed, can determine this functioning.

For now this applies to single particles. However, everything in the universe is made up of single particles. Through subsequent studies, we are discovering that consciousness can also intervene on aggregations of photons, atoms, molecules. Perhaps, in the future, we will discover that consciousness can intervene on whole biological organisms.

Perhaps we are talking about that phenomenon that in the extrasensory field is called psychokinesis?

Psychokinesis or telekinesis or Pk is a paranormal phenomenon. According to psychokinesis a living being is able to act on the environment that surrounds it, and can manipulate inanimate objects, through ways unknown to science.

Through psychokinesis it would be possible to move objects, bend metals, set in motion machines and perform many other actions. All this would be possible, with the sole power of the mind. Science does not recognize this possibility, and considers it a fantasy.

I add some philosophical considerations.

The official science has proved that by observing a single particle it is possible to influence its behavior. The observation collapses the particle on a given state.

Of course, everything that happens around us is the result of particles that collapse in certain states rather than in others. Moreover, it is true that we are the observers, so we are the ones who make them collapse.

Therefore, the sky appears blue to us and the peel of an apple appears red to us for the reason that, observing them, we ourselves determine that color.

It is true that we all see the same color, but not exactly the same. For this we can imagine having been designed with some common basic skills. It was necessary to maintain order in the creation. Who designed us did it with great wisdom, with great balance.

We are not destined to passively suffer what is happening around us. According to this project we are directors and builders of the universe that surrounds us.

Our senses have been designed to collapse reality just as it collapses.

There would be many other possibilities. For example we could see a red-green striped sky or a sea of black water. But these possibilities disappear when we "look" at the sky and the sea, and our gaze produces the known colors and transparencies.

So too the tastes we taste and the hardness we perceive to the touch are the result of the way in which the elementary particles involved in these phenomena collapse.

This confirms the most ancient thought of humanity. We are not mere products of the case. We are not prisoners in a universe that does not care about our insignificance. Rather, we live in a universe built to our measure. More: we ourselves contribute to building our universe.

A scientific nemesis

In this chapter we see how humanity is witnessing a scientific nemesis. A few centuries ago, some scientists removed the Earth from the center of the Universe. Today, other scientists are placing Man in the same position. We will approach the confirmation of this statement step by step.

Isaac Newton, English mathematician and astronomer, elaborated the laws of motion. Newton published the results of his research in 1687, in the volume "*Philosophiae Naturalis Principia Mathematica*". Every scholar, reading this book, understood that it would be possible to work out the trajectory of any projectile, in addition to the orbit of any celestial body.

In 1845, the French astronomer Urbain-Jean-Joseph Le Verrier calculated the position of a mysterious celestial body that was responsible for the irregular functioning of the orbit of Uranus. He did this by applying Newtonian principles.

In 1846 the German scientist Johann Gottfried Galle was able to observe the celestial body of Le Verrier. That body became the eighth planet of the solar system, with the name of Neptune..

This and other confirmations of the Newtonian theory suggested that, knowing the forces involved, it would have been possible to determine with absolute precision the motion of any celestial object.

In fact, it was possible to calculate the ephemeris tables. These tables contain the coordinates of the stars according to the course of time, ie they predict the positions that will be assumed by the celestial bodies in later times.

The enthusiasm was great, to the point that the French astronomer Pierre-Simon de Laplace (1749-1827) stated:

> "If the position and the momentum of a particle were known with precision in a given instant, then, knowing all the forces acting on the particle itself, its motion would be determined, in a univocal way, in all subsequent instants. This would be possible using the equations of mechanics ".

In simpler words, Laplace meant that if it had been possible to analyze the data relating to the position and velocity of all the molecules and atoms present in the universe, the consequence would have been staggering. In practice, using this knowledge, it would have been possible to establish the future movements of each celestial body. In practice, the fate of the universe could have been calculated.

Laplace's statement, despite its purely scientific features, had very relevant philosophical implications.

According to Laplace, if the situation of the universe was known at a given moment, it would have been possible to calculate all that had happened in the past and all that would have happened in the future.

Thus Laplace reinforced the idea, already widespread in scientific circles, of an absolutely mechanical universe, governed by chance and matter.

The whole universe was similar to a giant spring toy, a clockwork mechanism capable of performing dynamically preordained operations. But in this universe any activity resulting from free will would have been excluded.

Consequently, even man, as a matter of matter and therefore subject to the Newtonian rules, was a mechanical toy capable of moving only in the ways allowed by the gears. This "toy" could continue to move until the spring had discharged. He would have had no chance to change his destiny.

These conclusions confirmed the concept of "determinism", already in vogue since the 14th century. According to determinism, everything evolves with mechanical precision, regardless of man's desires and will.

After removing the planet Earth from the center of the universe, Man was also removed from the center of creation. The man became a creature generated by chance on a tiny planet located on the edge of the Galaxy.

Between the nineteenth and twentieth centuries the vision of the world changed radically, when physics and astronomy reached levels of knowledge that could not have been imagined in previous centuries.

Amazing coincidences

The task of science is to describe the universe through hypotheses and theories expressed in the form of universal laws. Very often the description uses mathematical formulas to reach a knowledge of reality in the most objective way possible.

In this way, proportions and ratios are calculated, which remain fixed under all conditions and are therefore defined as "constant". These values have an inestimable value. They are the very solid bases to be used for calculating all the laws that govern the universe.

Of course, the constants do not arise by chance and cannot be determined "one off" to support a theory. The constants must be confirmed as stable, fixed and invariable in infinite practical experiments. Only after this path are accepted by all physicists.

There are some ongoing controversies among scientists that some constant may vary over millions of years due to changes in the universe. For example, we hypothesize that the "universal gravitational constant" is decreasing as the universe ages.

However, this does not affect an observation that appears to be absolutely evident to everyone: if the physical constants had even imperceptibly different values, the universe would not exist or would be totally different from how we observe it.

That is, the universe is like this because the constants that regulate it are exactly like this, to the thousandth of a thousandth.

Consider the case of the constant that regulates the electromagnetic force in atoms, also called "fine structure constant" symbol "α". A small change in this constant would disrupt the relationships between the repulsive and attractive forces that are present among the elementary particles. In a universe with a different "α" value the Sun, its planets and any life form, including ourselves, would no longer exist. The "fine structure constant" is the value that regulates the relationships between the main physical constants of electromagnetism, ie the charge of the electron, the dielectric constant in a vacuum, the Planck constant and the speed of light. This constant is of great importance in string and multiverse theories.

The "Plank constant", commonly called "quantum", is represented by the symbol "**h**". The "quantum" is the smallest part of the elementary particles that make up matter: electrons, protons, neutrons and many others. The "how much" is indivisible. Its value is equal to $6.62606876 \times 10^{-34}$

Js. Just read this number to understand that it is an extremely small amount. Nevertheless it defines every part of the universe, from the size of the atoms to the force of nuclear reactions in the stars. Moreover, the light is made of "quanta".

The **"G"** symbol indicates the "gravitational attraction constant", (gravity constant) ie the proportionality coefficient in the universal gravitational law formulated at the end of the seventeenth century by Isaac Newton. The same constant also appears in the equation of the gravitational field of General Relativity. The constant **"G"** determines the force, proportional to the masses, with which it attracts any object, from stones to planets, stars or galaxies. The **"G"** constant is very small, it is equal to 6.67×10^{-11} N m² / kg².

The **"e"** symbol indicates the electron charge. The unit of electric charge is equal to $1.60219 * 10^{-19}$ coulombs. It is not divisible, there are no submultiples. Quarks that have a 1/3 or 2/3 charge are linked to other quarks in order to sum a whole unit or a multiple together.

Another constant that is of absolute importance is the speed of light in a vacuum, represented by the symbol **"c"**. Its value is equal to 299.792.458 m / s, often simplified in 300,000 kilometers per second. It is the speed that cannot be overcome in Einsteinian physics.

The four fundamental forces that govern nature are gravitational interaction, electromagnetic interaction, weak nuclear interaction and strong nuclear interaction. These forces depend on some of the constants mentioned, such as the speed of light, the universal gravitation constant, the Planck constant, the Hubble constant, the electric charge of the electron, the electron mass, and others.

But why do these constants really have those values? The answer may be implicit in another question: What would happen if the fundamental constants had different values?

We can do simulations, assigning to the constants slightly different values than the existing ones. In this way we can verify which kind of universe could result.

Well, any simulation shows that, by varying the constants, the conditions that allowed the development of life on Earth would not have been fulfilled.

Let's start from the extremely small. If the mass of the proton became greater or less than the mass of the neutron, all the atoms would become unstable. In practice, the universe would collapse into a cosmic jam.

If the hydrogen atoms contained protons and neutrons of different mass the result would be that these atoms would split into neutrons and neutrinos. The Sun and all the stars, without nuclear fuel, would be extinguished.

According to the simulations, if the strong nuclear force became minimally weaker, the only stable element in the universe would be hydrogen. Any other element would be absent. For example, there would be no carbon, which is the basis of our life.

We also consider the density of matter. If the density were greater, only black holes could be formed, not stars. Even if the density were smaller, the stars would not have formed.

If the force of gravity were little stronger than it is, the universe would be subject to a very rapid evolution and the stars would consume their fuel in a very short time.

If, on the other hand, the force of gravity were weaker than it is, matter would not be able to condense into nebulae, galaxies, stars and planets. The universe would be a chaotic space covered with fragments of material and gas.

In conclusion, it is clear that the physical laws that govern the universe could not be very different from what they are. If these laws were different, they would compromise the possibility of life as we know it.

Let us leave the cosmic space and evaluate other extraordinary coincidences that operate closer to us, at the level of the planet Earth.

Consider the distance of the Earth from the Sun. If it were less than a small percentage, like 5%, the oceans would boil. If, on the other hand, the Earth were 15% further away from the Sun, the whole planet would become a block of ice. By symmetry, the same things would happen if the Sun were slightly larger or smaller.

Carbon and oxygen atoms are almost equally present in biological organisms. This slight imbalance makes life possible. A different composition would create huge problems. For example, soils with excessive presence of oxygen would lose fertility because they would burn any carbon-based life.

The earth's orbit is eccentric, that is the non-circular but slightly elliptical. Furthermore, the Earth's axis is inclined. These two peculiarities contribute to maintaining a sufficiently stable climate, and allow the seasons to alternate. This makes agricultural crops possible.

For some years now, the search for Earth-like planets has begun. This research is focused on the so-called "habitability zone". It is a narrow band of space around the main star. Fortunately for us, the Earth is situated right in the small habitability zone that surrounds the Sun. If our orbit were more internal or more external, life on our planet could not exist as we know it. Thanks to the fact that the Earth is located in the habitable zone of the Sun we can have water in a liquid state.

The universe is intelligent. The soul exists.

Figure 17 - Brandon Carter, creator of the theory called "Anthropic principle". According to this theory the universe has been "constructed" as it is to allow the development of intelligent life.

Terrestrial biology is based on carbon, an atom with six protons. Carbon is not here by chance. This element was born during the formation of the universe through very complicated events. Thanks to these complex events, carbon has become the main constituent of our biology. It is impossible to say that this happened without a project. All the carbon that exists on Earth and on other planets was generated in the stars, during the formation process of the universe.

The process that led us to be here, to look, touch and shape nature with our hands and our eyes, began with the same origin as the universe.

After the Big Bang, the first element that appeared was hydrogen equipped with a single proton. Only 200 seconds after the Big Bang, from the fusion of pairs of hydrogen atoms, helium began to form which has two protons, and from the fusion of three hydrogen atoms lithium was born, which has three protons.

Continuing this process, beryllium was born from the fusion of two helium atoms, each with two protons. It has four protons.

The next step was the fusion of the beryllium atoms, endowed with four protons, with helium atoms that have two protons. This fusion led to the birth of the carbon atom that has six protons. However, these carbon atoms were unstable. Immediately after their formation these atoms disintegrated and again formed three helium atoms.

However, in the future, it was necessary to have stable carbon atoms to generate life. In particular, a planet, the Earth, needed stable carbon. The earth was not yet born, but stable carbon was in its future.

Incredibly, a process was developed in the stars to stabilize the carbon atoms. This happened when hydrogen was scarce in the stars and the temperature rose to about 100 million Kelvin. These conditions generated stable carbon.

All this still happens today, in the cosmos, but it is not enough. Another condition must occur: the carbon must leave the environment of the stars, where it is born, to invade the celestial bodies with temperature conditions more favorable to life, ie the planets.

Well, a providential mechanism solves this problem. When a star, at the end of its life cycle, becomes a supernova, it explodes and pours huge masses of matter into the universe, including carbon.

It took at least ten billion years for this whole process to be completed for the first time, after the birth of the universe, This means that the current age of the universe, about thirteen and a half billion years, is the most suitable to ensure that carbon-based biological life forms could develop on planet Earth, and probably also on other planets.

Ultimately we must recognize that the universe is a very delicate structure in which another equally delicate structure that is the planet Earth is inserted.

The universe and the Earth are born from millions, even billions of combinations that would have been possible. But only one condition has been fulfilled, so we are here.

We have won the lottery of life. It was precisely the only combination that could make our existence possible.

This statement was endorsed by eminent scientists, and is the basis of an interpretation of life in the cosmos called the "Anthropic principle"

According to the creators of the anthropic principle we can make an apparently banal statement. We exist and we are here to observe the Universe precisely because the universe has these particular characteristics.

But it is not enough, there is much more. Some say that the Universe is made this way because "an intelligence" wanted us to be here: that is, man is not a random product, but the initial and final objective. Man is a desired goal. The will that man existed wanted and produced the creation of our universe in the conformation suitable to host it.

The anthropic principle

For what was said in the previous chapter the probability that led to the birth of life, if due to chance, would be 1 on a number followed by such a quantity of zeros difficult to write in full.

Nevertheless, in traditional science dominated by determinism, man is considered a random zoological experiment, a secondary product of evolution.

On the basis of this assumption, there is no purpose in creation. Consequently, man is an aggregate of matter that does not imply any purpose. Also the human conscience is considered the product of particular molecular arrangements, which have been constituted, over millions of years, by random mutations and by the selection made by the environment.

According to the deterministic interpretation consciencnce, thought, intuitions and desires are waste products of chemical processing of the brain. All these spiritual motions have no correspondence with reality, indeed they are born and die in the cerebral convolutions. Positivist determinism holds that all these things are dreams, illusions, epiphenomena that disturb the machine's functioning. Man is a machine and everyone knows that machines do not dream and have no desires. However, man is recognized as having the ability to delude himself.

Obviously this is in disagreement with what was stated in the previous chapter. Why should the universe have been formed solely for the purpose of allowing the birth of the human being, if this being is little more than a mineral capable of moving?

It is clear that all the extraordinary consistency of the basic constants and the physical conditions of the planet cannot be considered random. The cosmos is based on order. It is true that in creation they could have constituted very different orders and relationships that would not allow the development of a life like ours. Probably this happens in other universes.

However, whatever order is established in a universe, it must be based on criteria that allow that universe to exist. The set of relations must necessarily be ordered to allow everything to work and to prevent the system from self-destructing.

Therefore, regardless of the presence of man, any universe in his order should be the result of a non-random process. Even a lifeless universe would need a project.

However, in our universe, the creative order wanted all the forces at stake to be such as to allow the development of our life.

This evidence has allowed many scientists to hypothesize and support the "anthropic principle".

If you ask around what people think about the anthropic principle, very few will be able to answer them. Some will have difficulty answering much simpler questions, like, "What is gravity?"

Gravity, and all the laws of nature, are natural norms that work even if we don't know how they do it. For example, almost nobody knows how our breathing works, yet we breathe for all the moments of our lives. Very rarely we worry about knowing how to do it, unless we are medical students.

The anthropic principle is something of the same type. If it didn't exist, we would live equally well without worries.

The fact that the anthropic principle exists has a completely philosophical importance, such as the existence of God or the fact that the Earth revolves around itself.

Whether we believe it or not, they are things that happen anyway. Maybe that's why we don't care about it at all. We believe that, however, these things do not change our lives.

In fact it would not be absolutely so, because if the Earth did not rotate on itself many things would change in our physical existence. Similarly, if God did not exist many things would change on the spiritual level and in the final outcome of our existence.

What drives many to investigate the great scientific, philosophical and spiritual themes is a certain flame that in some burns stronger than in others.

This flame is the curiosity, the desire to know, the ambition to discover the mechanics of the gears to modify to our advantage the operations of what surrounds us, under and above the sky.

Whether we live for the universe, or that we live in the universe, or that the universe lives for us, changes little in our daily lives and is enough to make the topic completely irrelevant for many.

But many others, like you who are reading this book, wish to investigate and know, because knowledge has always been the spring of human evolution. Without knowledge we would still be here to eat raw lizards after capturing them by throwing stones.

The anthropic principle is a theory that has not yet been confirmed (it would be very difficult to do so). However, this theory is affecting many of the most enlightened scientists.

Birth and evolution of the anthropic principle

Paul Dirac, physicist and mathematician, winner of the Nobel Prize for Physics in 1933, is one of the founders of quantum mechanics. He was born in Bristol, in the United Kingdom, in 1902. Dirac was the first to notice the existence of strange affinities between very different physical quantities.

In fact, in the 1930s Dirac calculated a strange equality. The square root of the estimated number of particles present in the universe is equal to the ratio between the electromagnetic force and the gravitational force existing between two protons. Dirac drew the conclusion that this relationship is not constant, but varies on cosmological times.

In the late 1950s Robert Dicke, another US experimental physicist, confirmed the surprising coincidence found by Dirac. Dicke stated that the equality of the two values was more evident in the first phase of the evolution of the stars. At that time there was a particular abundance of carbon, which is the fundamental constituent of living organisms.

Therefore, the coincidence found by Dirac was undoubtedly associated with the evolutionary processes responsible for the appearance of living forms based on carbon chemistry. In 1957 Dicke expressed his thoughts with these words:

> "The current age of the universe is not accidental but is conditioned by biological factors. Any change in the values of the fundamental constants of physics would prevent man from being here to measure them".

In fact, this was the first statement of the weak anthropic principle. It was an unconscious statement, because the anthropic principle was not yet known. Therefore, Dicke's remark was greeted with indifference. His idea was not subject to unfavorable prejudices. However, prejudices rained abundantly when the anthropic principle was elaborated and published, that is when the principle was understood in all its implications.

The theory was first enunciated, in an official way, in 1973 by the Australian physicist Brandon Carter (figure 17). The first version of the theory evolved into various interpretations: the "weak principle", (weak anthropic principle), the "strong principle", (strong anthropic principle) the "ultimate principle" (ultimate anthropic principle) and the "participatory principle". (participatory anthropic principle)

In the weak principle the theory is of a disarming obviousness, consequently few dispute it. This version states that the universe in which we live, in fact, allows life as we know it. This statement comes from a profound knowledge of the laws of nature. These laws establish that life is allowed thanks to innumerable fortunate coincidences, all indispensable. If one coincidence were not true or different, life would not exist.

The exposition of the weak principle contains this statement:

"The values of all physical and cosmological quantities are not equally probable. The values of these constants respond to the condition that places must exist in which carbon-based life can evolve. Furthermore, these values respond to the condition that the Universe is old enough to give rise to carbon-based life forms ".

Later Brandon Carter theorized the strong anthropic principle. In 1986 John Barrow and Frank Tipler, in the two-handed book "The Anthropic Cosmological Principle", analyzed the principle according to the "strong" version. In the strong principle it is stated that the universe "*must*" possess those properties that allow life to develop within it.

That "must" shifts the focus from purely scientific to philosophical or metaphysical. In fact, the verb "must" presupposes the existence of an Entity that expresses and exercises its will in the creation of the universe. This statement is very indigestible for positivist science.

However, the strong formulation of the anthropic principle describes a new relationship between the universe and man. An ancient dignity that had been taken away is recovered and returned to the man. The strong anthropic principle removes the human being from the marginal position in which he was relegated, along with his planet. Naturally, the Earth never returns to the center of the Universe. The central position is occupied by Man, around which and for which the universe exists.

The strong anthropic principle has the double merit of restoring prestige to the human being and of "illuminating" science with new nobility, removing it from the mechanistic greyness of the Enlightenment. (When we say the meaning of words!).

Barrow and Tipler have been fiercely criticized because in their book they propose a third typology of anthropic principle, which is added to the two theorized by Carter. The two authors propose the "ultimate anthropic principle". With this further theory they want to better explain the incredible coincidences that allow the existence of our universe and intelligent life.

In the "ultimate principle" Barrow and Tipler start from the postulate that, in the case of infinitesimal variations of the values of the fundamental cosmological constants, the existence of the universe as we know it would be lost. Considering this, they conclude that we cannot study the current structure of the universe without taking into account our physical needs. The statement of the last anthropic principle states:

> "Intelligent information processing in the universe must necessarily develop. This intelligence, once appeared, will never die out ".

At first, the famous astrophysicist Stephen Hawking expressed doubts. He stated that the existence of other galaxies and the large-scale homogeneity of the universe could be in contrast to the strong anthropic principle.

Later, however, when he devoted himself to the development of the Theory M, Haking changed his mind and became a strong supporter of the anthropic principle. He inserted in many of his equations a variable related to this theory.

Finally, another famous American physicist, John Archibald Wheeler, suggested the theory of the "participatory anthropic principle". This is an alternative version of the strong anthropic principle. Wheeler describes his thought thus:

> "The universe must be such as to allow the creation of observers within it at a given stage of its existence. Observers

are necessary for the existence of the universe, as they are necessary for its knowledge. Thus the observers of a universe actively participate in its very existence ".

The participatory principle is a variant of the strong principle. This principle reverses reasoning and argues that the universe exists because we exist.

The American astronomer Hubert Reeves describes the anthropic principle as follows:

> "The anthropic principle can be formulated more or less in the following way: since there is an observer, the universe has the properties necessary to generate it.
> Cosmology must take into account the existence of the cosmologist. These questions would not have been asked in a universe that had not had these properties ".

"The Melancholy of Haruhi Suzumiya" is a series of "light novels" that was written by Nagaru Tanigawa and illustrated by Noizi Itō. In 2003 it became a series of films broadcast all over the world. In this Japanese series the concept of anthropic principle is described as follows:

> "According to this theory we observe the universe, and for this reason the universe exists. Humanity, the only intelligent life on our planet, has discovered the laws of physics and their constants and has been able to describe how the universe is made. In this way the awareness of the existence of creation and the act of observing it end up coinciding ".

All the latest quotations refer to a concept that may seem confusing at the moment, that of "observer". The role of the observer is framed in the context of quantum physics and is absolutely crucial. I will talk about it extensively in the next chapters

Is man really at the center of the universe?

All science, from Kepler onwards, has drastically reduced the ambitions of those who placed the Earth at the center of the universe.

The anthropic principle replaces the centrality of the Earth with that of Man. All creation exists as a function of the development of life, especially of intelligent life.

It is easy for all of us to identify "intelligent life" with "man". When we talk about "man", we mean "the inhabitant of planet Earth".

Over the centuries, this ambition has undergone infinite downsizing. Despite this we do not give up occupying the central role that we feel is our due to a hypothetical superior right.

Since we can no longer put our planet at the center of the universe, we occupy that place ourselves. Unfortunately, this attempt is also destined to be mortified.

The attempt would certainly be legitimate if we lived alone in the universe. Instead, in recent decades a new science is working hard to disappoint our hopes of universal supremacy.

This science is called exobiology.

Exobiology is a field of biology that studies the possibility of extraterrestrial life and the nature of this life. Exobiology is currently a speculative sector, but for most scientists it is a valid field of scientific exploration.

Computer simulations were performed on the possible existence of life processes in environments outside the Earth. These simulations have indicated the possible existence of life forms similar to ours or even alternatives. For example, there may be life forms based on silicon rather than carbon.

In the last century most scientists exchanged ironic giggles while hearing about extraterrestrials. Those times are far away. Currently there are research projects of life in space financed with millions of dollars by States and Organizations of various types. We can first cite the SETI astronomical radio listening project, experimentally begun in 1960.

SETI (Search for Extra-Terrestrial Intelligence) was officially launched in 1974 in Mountain View, California. This is a program dedicated to the search for extraterrestrial intelligent life. SETI takes care of listening and sending radio signals to other civilizations.

In the 1960s a method was created to measure the possibility of the existence of planets inhabited by other civilizations. This is the "Drake equation". The method is named after its creator, Frank Drake, an American radio astronomer.

The Drake equation, often also called the "Green Bank formula", was formulated in 1961. It represents the attempt to estimate the number of extraterrestrial civilizations existing in our galaxy, the Milky Way.

Unfortunately the results are uncertain due to the lack of any reference point. The only useful reference is the existence of life on Earth. But this is already a good starting point. Why should life exist only on a planet among billions?

Applying the formula, in the light of current astronomical knowledge, the extraterrestrial civilizations that would be able to communicate with us would be thousands only in the Milky Way.

The formula of the Drake equation is the following:

$$N = R * Fp * Ne * Fl * Fi * L$$

where is it:

N is the final result, that is the number of extraterrestrial civilizations present today in our Galaxy.

R is the average annual rate at which new stars are formed in the Milky Way.

Fp is the percentage of stars that potentially have planets. It indicates how many planets, among those revolving around a Sun, would be in a position to host life forms.

Fl is the percentage of Ne type planets on which life has actually developed.

Fi is the percentage of the planets Fl on which intelligent beings would have evolved.

Fc is the percentage of extraterrestrial civilizations able to communicate.

L is the estimate of the duration of these evolved civilizations, before their extinction.

Since 1961 many values have changed in a favorable sense. The Kepler satellite, after nine years of exploration in our vicinity, has discovered about 2600 probably habitable planets. This is a much higher percentage than estimated in the formula.

Recently, Italian researchers have confirmed the existence of water on Mars. This discovery increases optimism about the possibility of life, past or future, on seemingly uninhabited planets.

The Italian Claudio Maccone, is an Italian SETI astronomer, space scientist and mathematician, awarded the "Giordano Bruno Award" in 2002.

Maccone has updated the values in the Drake formula according to the recent parameters accepted by SETI.

In this way it was possible to formulate a more precise estimate of the potential extraterrestrial civilizations. Claudio Maccone has established that the hypothetical number is between 0 and 15,785, with an approximate average of 4,590.

There is 75% of the chances that these civilizations are at a distance between 1,361 and 3,979 light years.

However, this is an enormous distance, which seems to exclude any possibility of communication.

Intelligence cooperation

Scientific research has accustomed us to incredible progress and to science fiction hypotheses. Often these hypotheses become everyday reality in a few decades or even in a few years.

Quantum physics, with entanglement experiments, has shown that elementary particles can communicate without time and space constraints.

The hypothesis of "black holes" and the "string theory" are practically still virgin fields of study. From here epochal revolutions could arise in the development of communications and in the possibility of traveling through wormholes (space tunnels).

Traveling through wormholes it is possible to overcome the speed of light, taking advantage of the curvature of space. The laws of relativity also make time travel possible.

We can hypothesize that within two or three generations, our descendants will come to know other civilizations. Of course, this can only happen if the man does not self-destruct before it happens.

What will happen when we meet other civilizations? No one knows.

Certainly humanity has been made up of explorers and pioneers, ever since the homo sapiens invaded the territories of the Neanderthals. This propensity was confirmed when the navigators, traveling on tiny boats, went beyond the waters of unknown oceans.

The declared purpose of the explorations was the desire to export civilization or the Gospel. Unfortunately, there was also an undeclared purpose. This purpose inevitably led to robbery and exploitation. All the explorations, in reality, were financed to achieve economic advantages.

Fortunately, time and revolutions made it possible to transform the exploited populations into independent communities.

There are territories that were initially exploited by the European powers. We can quote entire continents, such as North America or India. Today these are independent nations. Their populations are no longer considered inferior, because they contribute to the development of a civilization increasingly oriented towards progress. Today's civilization is inspired by values of friendship and fraternity, even though these are often nominal values.

With what spirit will the terrestrial man approach other extraterrestrial civilizations? Will he do it with the spirit of robbery, which has always been congenial to him? And these civilizations, many of which will certainly be more advanced, how will they approach us?

We can make a forecast. In the centuries of geographical discoveries, the objective was to search for gold, silver and other precious materials. Today, however, the good that can affect both extra-terrestrial civilizations and our civilization is another: knowledge.

Knowledge is a raw material that does not need giant spaceships to be transported. Furthermore, knowledge does not belong to a single individual but to entire systems that work together to maintain and increase it. As a result, knowledge cannot be extorted by violence against individuals.

The exchange of knowledge requires a consenting and conscious collaboration. This is the kind of collaboration that will probably come about with extraterrestrial civilizations.

Surely it will no longer happen that on a journey of space exploration someone can land on the planet X giving away mirrors and necklaces to the inhabitants. Of course, not even we would accept such gleaming merchandise if some extraterrestrial landed in one of our cities.

Probably, considering the immense distances, future exchanges can take place only in symbolic form, through the transmission of thought or with the help of new technologies all to be developed.

It is evident that the different size of the planets, the different conformation of the atmosphere and the different conditions of pressure, gravity, heat, circadian and seasonal cycles will make the physical life of other species on the earth's surface impossible. We too would have difficulties to adapt ourselves to living on other planets. We are becoming aware of what serious physical problems the astronauts involve, in the very short trips made in our solar system.

A cycle of physical adaptation to the conditions present on other planets can take place only through hundreds or thousands of generations. At that point it would no longer be possible to distinguish between "us" and "them".

Instead, scientific information can travel smoothly from one planet to another.

Ultimately it is likely that, when contacts are established with other civilizations, man will make his gregarious character prevail over the ambitions of oppression and robbery.

Moreover, man tends to be gregarious. We live the gregariousness so extensively that we hardly notice it anymore. We create families, we organize work groups, committees, councils and assemblies, we provide ourselves with regulations of our condominiums, we establish laws in cities and nations. Peoples and supranational institutions collaborate in research against diseases, in the development of new technologies, in the diffusion of values of civilization as the protection of the weakest.

Perhaps, if we come into contact with alien civilizations, these civilizations will help us improve by teaching us fraternity and cosmic collaboration.

At that point the difficulty arising from the anthropic principle will be solved. Was the universe born to favor only the life of man on Earth?

We will probably understand that the Universe was born to favor any intelligence, wherever it is. The inhabitants of the Earth and those of an incalculable number of other planets will establish collaborative forms that will guide civil progress towards goals that are currently unthinkable.

We cannot help but see in this the project of a cosmic Mind. This project develops through processes of synchronicity aimed at a very specific goal: the triumph of intelligence. The triumphant intelligence will be able to understand, finally, the Mind that wanted and organized this project.

Synchronicity is a process that can affect individuals. Synchronicities, through strange coincidences, dreams, apparently disconnected events, guide us towards a process of psychic improvement.

But synchronicities can also affect groups, communities, peoples. They have interested entire civilizations. For example, Joseph Cambray speaks of a synchronicity that, in a few years, made Greek democracy blossom from nothing.

A synchronicity is affecting all humanity that has lived for millions of years in the brute state of the stone age. In the last ten thousand years, suddenly, the history of humanity has literally exploded passing from the stone age to that of space travel.

Another synchronicity is working so that all the civilizations of the universe come to know and understand each other, exchanging knowledge with each other. At the end of this synchronic process all intelligent beings will be transformed from mere mortals to new gods.

Creatio ab nihilo

"All matter originates and exists only by virtue of a force that makes the particles of an atom vibrate and holds together the tiny solar system of the atom.
We must suppose the existence of a mind conscious and intelligent behind this strength. This mind is the matrix of all matter. "

(Max Planck, German physicist, 1858-1947)

What evidence do we have on the intelligence of the "Cosmic Matrix"?

"*Ex nihilo nihil fit*" is a way of speaking of the Latin language that can be translated as "From nothing comes nothing". The Latin poet and philosopher Lucretius expressed this principle in the first book of the "*De rerum natura*" (I, 149-150):

> "We can begin by saying that nothing ever comes out of nothing, by divine will."

Lucretius was a follower of the atomistic philosophy of Democritus. Democritus maintained that matter, in the form of atoms, is eternal. Even the French chemist and biologist Antoine-Laurent de Lavoisier, who lived several centuries later, maintained a similar concept:

> "Nothing is created and nothing is destroyed, but everything is transformed".

This statement supported the law of conservation of the mass. Subsequently Einstein's confirmation arrived, expressed in the most famous formula of our times: $E = mc^2$.

With this formula it is confirmed that the mass can be transformed into energy and vice versa. The principle that the sum of energy and mass in the universe is constant remains valid.

Figure 15 proposes a "Cosmic Matrix" in which an infinite number of universes is generated. This matrix can be infinite, because it has the same characteristics as thought. The Cosmic Matrix could host an infinite number of universes besides ours. We ignore the existence of these universes, and we will probably continue to ignore it for the eternity of time. Of course, time is also our convention.

I have called this matrix "Universal Mind", a neutral name that everyone can freely transfer into other philosophical or theological concepts, according to his culture and his beliefs.

The question we are now asking ourselves is whether this universal Mind is limited to generating random, inconsistent and meaningless thoughts, or if instead its thoughts are orderly and coherent, that is, intelligent.

Based on the characteristics of a "mind", as we understand it, both possibilities coexist. For example it also happens to our mind to develop over time a project that is consistent with our intentions and our creativity. In the same way our mind can get lost in suggestions, imaginations, flashes of light that come on and immediately disappear, meaningless reflections and deductions on topics that pass very fast and elusive.

This is certainly true in dreams, when the mind is free from the conditioning of reality and can wander aimlessly through the surreal territories. These escapes in irrationality happen and involve us even though we are intelligent beings.

However, most of the brain elaborations of our mind are devoted to the planning and implementation of coherent projects.

In this sense, is the Universal Mind intelligent?

The Universal Mind designs universes. His thoughts are aimed at creating universes. His projects are walnut shells generated in infinite quantities within an infinite space of thought.

We do not know if all universes are intelligently designed. That is, we do not know if all the universes are created with the aim of making them develop and evolve in an orderly way towards a finalized goal.

Certainly, however, our universe has this purpose.

From what has been said above, and for all the reasons that justify the theory of the anthropic principle, our universe was born from a project that provided from the beginning what would be the constants and natural forces that would lead to the development of life.

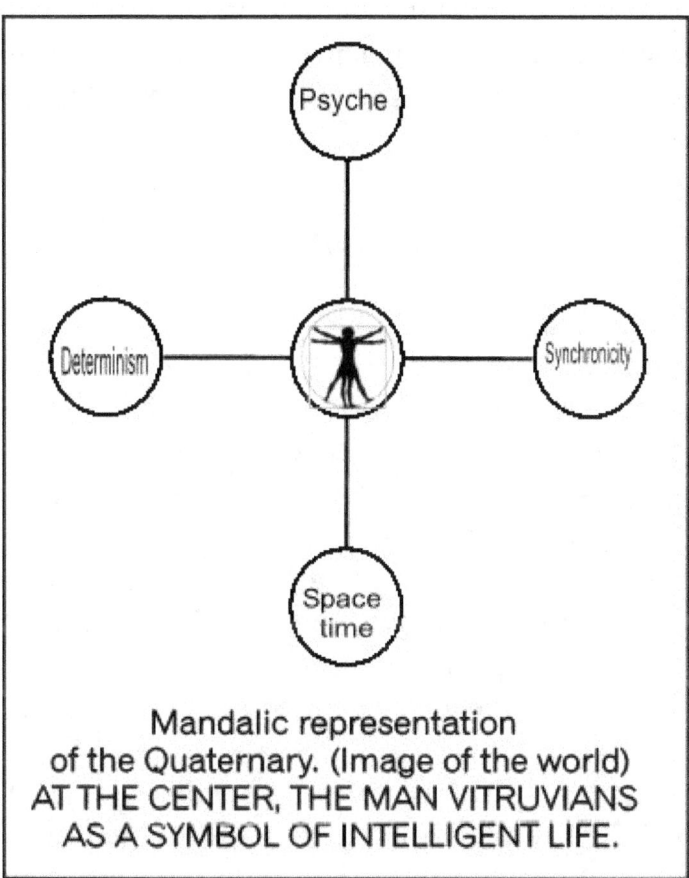

Figure 18 - The psycho-physical diagram of Jung-Pauli represented in the form of a mandala. The symbol of intelligent life is placed in the center.

We are here, because the "cosmic Mind" wanted us to be here. With immense intelligence the Mind has set up the optimal conditions to make this happen.

But does the universe made of matter really exist?

The Big Bang theory has always had a weak point in the fact that the universe could have started from a singularity. It is very difficult to imagine that at the origin of the world all the mass and energy were concentrated in an infinitesimal point.

Quantum cosmology scholars have come up with a truly surprising proposal to solve this problem. This is the theory called "total zero energy universe".

This theory is based on the hypothesis that the total energy of the universe is equal to zero. In practice, the positive energy due to the matter would be exactly balanced by the negative gravitational energy. Consequently the energy is canceled for a simple sum. If we add +1 to -1, the total is zero.

In 1973, the American physicist Edward Tyron published an article in the scientific journal "Nature". Tyron argues that the entire universe emerged from quantum fluctuations in vacuum. These fluctuations would be able to create pairs of particles and antiparticles.

Stephen Hawking writes this in one of his articles:

> "In the region of the universe that we can observe there are something like 10^{80} particles of matter. Where did they come from? The answer is that, in quantum theory, particles can be created in the form of pairs consisting of a particle and its antiparticle. This happens starting from energy.
>
> At this point, however, another problem emerges. Where did this energy originate? The answer is that the total energy of the universe is exactly zero.
>
> The matter of the universe is made up of positive energy. However it must be kept in mind that all matter is attracted by the force of gravity. Two pieces of material placed close to each other have less energy than two identical pieces placed at a great distance. This happens because, in the case of the two neighboring pieces, energy must be spent to keep them separate against the gravitational force which tends to bring them closer

together. Instead, in the case of the two distant pieces, the necessary energy is negligible. So, in a sense, the gravitational field has negative energy.

If we take the whole universe, we can show that the total negative gravitational energy is exactly equal to the total positive gravitational energy. The result is that the two energy cancel each other out, so that the total energy of the universe is zero ".

At this point Fracastorio's statement in the "Cena delle ceneri" becomes current:

"Nullibi ergo erit mundis. Omne erit in nihilo. "
(Then the world will not exist. Everything will be nothing, equal to zero).

It's amazing how the character created by Giordano Bruno could have such an appropriate intuition.

According to the theory of "Total zero energy universe" the existing mass in the universe, which has a positive sign, is exactly equal to its gravitational energy, which is negative.

This produces a disconcerting effect, which can be summarized in two points and a consequence:

- The expansion of the universe generates an increase in negative gravity.

- The negative force of gravity generates an increase in the mass of positive value. This happens to preserve the equality of the two gravities.

The consequence is this:

- The mass is spontaneously created at the expense of expanding the universe and increasing the force of negative gravity..

In conclusion, Edward Tyron argues the possibility that the creation of the universe starts from the particular quantum fluctuations of the initial vacuum. These fluctuations, as already mentioned, generate particle-antiparticle pairs.

Furthermore these fluctuations, starting from a microscopic region, would have originated small areas that are not homogeneous and unstable from the point of view of the mass / gravity ratio.

These areas of instability, magnified by a process called inflation, have given rise to increasingly larger structures up to galaxies and galaxy clusters.

In other words, if we assume the original existence of the quantum void, the formation "ex nihilo" of the universe becomes possible and derives from natural laws.

The question remains who would have written these laws.

The problem is also another, and we cannot solve it here. If matter and energy in the universe are zero, then matter and energy do not exist.

So are we talking about a universe or a ghost?

Is there anything that has been created "ex nihilo", or is it all a great illusion? Perhaps all that we identify as "real" is only a big dream. The fact remains that we dream.

Non locality, entanglement

From the point of view of common sense, quantum electrodynamics describes an absurd nature. However, it is in perfect agreement with the experimental data. I therefore hope that you will be able to accept Nature for what it is: absurd.

(Richard Feynman, American physicist)

Einstein and the locality

If there is something that can upset a Swiss, that is a person born in the land who is by definition the home of watches, it is the lack of precision. If then this person is engaged in a methodical work which may be that of a patent office clerk, then we can understand his disappointment when things do not work as they should. This is more true if this person, in his spare time, is also a mathematician.

We are talking about Albert Einstein. We are talking about Albert Einstein. Throughout his scientific career, Einstein only encountered one thing that made him irritate: "quantum indeterminacy". Einstein wholeheartedly detested the lack of discipline of elementary particles and their characteristic of elusive ambiguity. Einstein detested the fact that elementary particles refuse to make their position in space and their speed known at the same time.

All this was a slap, indeed, a sneer at the good and solid rules of Newtonian physics, on which Einstein rested his most known theory, that of relativity.

Newtonian laws are based on the "principle of causality", also known as determinism. The universe is made of matter. In the realm of matter

The universe is intelligent. The soul exists.

nothing happens by chance, everything happens as a result of something that happened previously. Matter attracts and repels, collides, moves or remains immobile. In these movements, matter pays a price with a currency called energy.

Only a certain energy, applied to the matter, gives rise to an action or a chain of actions.

Imagine a football game. There is a well placed ball on the spot in the penalty area, waiting for a player to throw it towards the goal. Do you think that the ball will launch itself towards the opponent's goal without receiving a player's kick?

Imagine a golfer who prepares to hit the ball to throw it towards the hole. That player is apparently impassive, but his soul is involved in hundreds of mental elaborations. He must calculate with the utmost precision what force and what angle he must give the ball to direct it towards the objective.

In fact, nothing will happen by chance. The ball will reach exactly the point corresponding to the push received. The success of the launch depends only on two factors. The first factor is the accuracy of the calculations made by the launcher. The second factor is the player's ability to transfer the calculations to his arm. Imagine that a ball, based on the calculations, reaches up to a millimeter from the edge of the hole. It will never happen that the ball decides, on its own initiative, to move a little further. Not even the cries and solicitations of the public can push the ball a millimeter further.

We know how to calibrate our energy to achieve the desired results, because we know that objects respond with absolute precision to our "commands".

For this we can launch probes in space and we can make them land precisely in the fixed places, whether they are on the Moon or on Mars or elsewhere.

Recently the European Space Agency, with the Rosetta mission, launched a space probe exactly on a moving comet. This comet, known as 67P / Churyumov-Gerasimenko, is just a tiny stone with a core 3 kilometers in diameter, which travels thousands of miles away in space. We hit it with absolute precision.

Is causality the basis of all things?

In 1950, Alan Turing wrote this in his book "*Calculating machines and intelligence*":

> "If we move a single electron to a billionth of a centimeter, this can produce the difference between two very different events. For example, a year later that move may result in the killing of a man due to an avalanche, or his salvation."

The universe is intelligent. The soul exists.

In 1972 Edward Lorentz gave a lecture entitled *"Can a butterfly's wing beat in Brazil cause a tornado in Texas?"*

In today's scientific world, everything can be weighed, measured and determined in the laboratory. This world is placed in a dimension where time only walks forward.

In this world made only of matter, the answer to Lorentz's question could be "yes".

In fact, from the point of view of Newtonian physics, every event can be predicted as long as it can be measured. To make the measurement possible, the parts in play must have a weight, a size and a place in the space. That is, the parts to be measured must be matter or time. Time is also measurable and predictable.

On the fact that a butterfly in Brazil can cause a tornado in Texas, we may have doubts.

If this could happen, it would only be a very indirect consequence. However, this possibility provides so many variables that it cannot be calculated even with the most powerful computers.

In physics there is the principle of "locality", according to which distant objects cannot have instantaneous influence on one another. An object is directly influenced only by a force placed in its immediate vicinity. It is necessary to take into account the weakening of the force of gravity with increasing distance. Furthermore, a signal sent in any way to a distant object needs time to overcome the distance, and the speed cannot exceed 300,000 kilometers per second, ie the speed of light.

Einstein was very concerned when signals of the existence of "non-local" relations between elementary particles began to arrive from quantum physics.

The problems that puzzled him most were those posed by Heisenberg's "uncertainty principle". This principle, enunciated in 1927 by Werner Heisenberg, represents a pivotal concept of quantum mechanics and constitutes an irreparable break with respect to the laws of classical mechanics.

Heisenberg showed that it is not possible to know simultaneously and precisely two "conjugate variables". For example it is not possible to know at the same time the precise position of a particle and its momentum, or speed.

This contrasted markedly with the demands of classical physics. Remember that with classical physics we can calculate anything if we know the starting values. To calculate the trajectory of a space capsule or a billiard ball I need to know exactly where it is at the beginning, which push it receives and at what speed it will move.

In quantum physics, these values are never available simultaneously.

If we measure the position and the momentum of a particle at the same time, the values obtained are absolutely uncertain. This uncertainty does not derive from measurement techniques, but is the consequence of quantum reality which is a "probabilistic reality".

The concept of probabilism is clear if we return to consider the experiment of the double slit: a photon thrown against the barrier with two slits has the probability of crossing one or the other, therefore it crosses them both.

Only the observer can collapse the different probabilistic states in a single point. In an unobserved situation, all probabilistic states exist. Speaking of quantum probabilism, Einstein uttered the famous phrase:

> "It is difficult to take a look at the cards that God has in his hand, but I cannot believe even for a moment that God plays dice".

Quantum entanglement

Quantum reality does not meet the criteria of local classical physics, therefore it is called "non-local". In the non-local domain there are events and demonstrations that do not suffer the typical restrictions of the local domain. Non-local domain events are not limited by time or distance. In this domain there are no "yesterday and today" nor "before and after". There is only "now and always". Likewise there are no "high and low", "near and far", but only "here and everywhere".

The most obvious confirmation of the existence of the non-local domain is given by one of the most famous experiments of quantum physics. This is the experiment carried out in 1982 under the guidance of Alain Aspect, a French researcher. This experiment confirmed the quantum entanglement theory and put an end to a very long period of protests. The main opposing theses were supported on the one hand by Niels Bohr, director of the study group called "Copenhagen School" and on the other hand by Albert Einstein.

The most surprising and intriguing feature of subatomic particles is the ability to instantly exchange information between them. In practice, information does not pass through a physical space to connect the two particles, that is, they do not make a path between one and the other particle. The non-local level is purely psychic. In the non-local level, exchanging information is like exchanging thoughts.

Entanglement is an English term meaning "weaving". This term represents the intertwining that is established between two "correlated particles" that is born together. Today the "entanglement" experiments do not concern only two particles. Quantum entanglement can also be achieved among millions of related particles in the laboratory. We cannot help but notice that if we consider the universe as a large laboratory, then the Big Bang, in its creative explosion, correlated all the existing particles.

The problem that besets classical physics is not the fact that a plot is established. The real problem is that this interweaving distorts all the laws of classical physics, that is the pillars on which modern science rests.

Classical physics establishes some things, including:

- The reality is causal and mechanistic: every action is the reaction deriving from a previous action and is the cause of subsequent actions.
- The speed of light limit cannot be exceeded.
- Each force (gravitational, magnetic, etc.) decreases as a function of distance.
- The arrow of time establishes a rigid hierarchy in the evolution of each event. What happens first is the cause of what happens afterwards and the opposite can never be true.

At the level of elementary particles none of these rules is worth more.

Let's start from the fact that the two related particles can be obtained in various ways, but in any case they have opposite "spin". One particle has "negative half" spin and the other has "positive half" spin. Spin is a property similar to the sense of rotation (right-handed or left-handed).

If one of the two particles reverses the spin, the other inverts it, not immediately but simultaneously. It doesn't matter how far the two particles are in the universe. So:

- The speed of light limit is no longer valid.
- The principle according to which forces weaken in function of distance is no longer valid.
- Since there is no time difference between the inversion of the two particles, the arrow of time is no longer valid and there is no causality.

The first practical confirmation occurred in the experiment carried out by Alain Aspect in 1980-1982. Later the experiment was confirmed hundreds or maybe thousands of times.

Everything is one in the non-local dimension

From this experiment a question arises that, at the moment, has no answers.

How does a particle, even placed at an astronomical distance, know that the other is changing the spin?

We can imagine that the particle B notices the change of the particle A AFTER that has happened. It is not so. In fact the particle B knows this at the same time, and the two particles simultaneously change the spin.

The two particles behave as if they were a single particle, that is, as if they were united in the same place.

What information did the two particles coordinate?

Through which field did the information travel to coordinate the two particles?

It is necessary to hypothesize a field, that is an imaginary space, which is not made of matter but only of energy and information. In fact this is a psychic space.

Is it perhaps the same space that Plato called "World of ideas" and that Carl Jung later called "Collective Unconscious"?

It is the space called non locality, because it cannot be placed anywhere, but it is everywhere. It connects the whole universe, so that every part of the universe is immersed in this level of energy and information. The entire universe contains only one energy and one unique information. The whole universe is one.

In this nonlocal level, where there is neither space nor time, all the information of the universe pervades our consciousness

It is the information that Jung called "archetypes". Synchronicities also arise in non-locality. The synchrionicities flow towards our conscience and generate all the curious coincidences of which we are protagonists, the presentiments, the spiritual intuitions. Synchronicities are windows open to spaces of the spirit.

The soul exists

Only the traveler who has wandered in his infinite inner world can approach the Soul. Thus he will discover that for years he has done nothing but look for the Soul, since the soul is behind and within everything.
(Carl Gustav Jung)

You are a little soul carrying a corpse around
(Epictetus, Greek philosopher)

The aggregation of matter

All the matter in the universe is made up of particles. The way in which matter appears to us is determined by how the particles are arranged, which are linked by attractive forces. The mutual attraction of the particles is contrasted because the particles themselves are in a state of perpetual agitation. The state of aggregation of matter depends on the resultant of these two opposite tendencies: attraction and agitation.

There are mainly three states of matter: solid, liquid and gaseous.

Solid state materials have their own shape and volume. In solids the molecules are mutually linked by intense forces and occupy positions that are on average fixed with respect to each other. The rigid structure of solid matter derives from the orderly and compact arrangement of the particles.

Even the liquid materials have their own volume, but take on the shape of the container that contains them. The particles of liquids can flow over each other, because their kinetic energy manages to overcome, in part, the attractive forces.

In the gaseous state the particles are distant from each other, and are in a state of disorder. They have no volume of their own. They are free from obstacles and tend to expand occupying all the available space. In the gases the forces of attraction between the single molecules are weak.

The state of aggregation is not a fixed characteristic in a substance: for example, water can assume the solid state (ice), liquid or gaseous (water vapor). Each substance can change its state. When this happens, the substance absorbs or releases energy in the form of heat.

The atoms that constitute the matter are ageless; they can pass from one substance to another, from one body to another and from one organism to another.

Who says "We are children of the stars" says a great truth. The atoms of our body existed long before us. At our death these atoms will be recycled into other manifestations of matter, biological or not.

During our lifetime, we certainly breathed at least one air molecule breathed by famous historical figures like Tutankhamun or Marylin Monroe.

The great variety of shapes and colors with which the material appears to our eyes is due to the fact that the atoms can join together in many ways and can form larger and more complex structures.

The universe is intelligent. The soul exists.

Figure 19 - Some scholars who have contributed to a spiritual and not exclusively materialistic view of science.

The matter aggregates into coherent and finalized forms

Atoms aggregate to form molecules. The molecules aggregate to form bodies of all kinds, from "Amoeba Proteus" to the Andromeda galaxy. The first observation that arises spontaneously is this: both the extremely small amoeba and the extremely large Andromeda contain a wonderful order that makes their existence possible.

If amoeba had no way to feed and reproduce, and if any basic value of the galaxy were not exactly as it is, neither of them would exist.

All this is trivially taken for granted, but the question no one knows how to give a coherent answer is: why does matter come together exactly like this?

Classical physics provides an apparently elementary answer: matter aggregates in this way because there are laws that spontaneously produce these aggregations.

This answer does nothing but move the problem further: why do these laws exist and not others?

Speaking of the anthropic principle we recalled the first steps of the evolution of the universe.

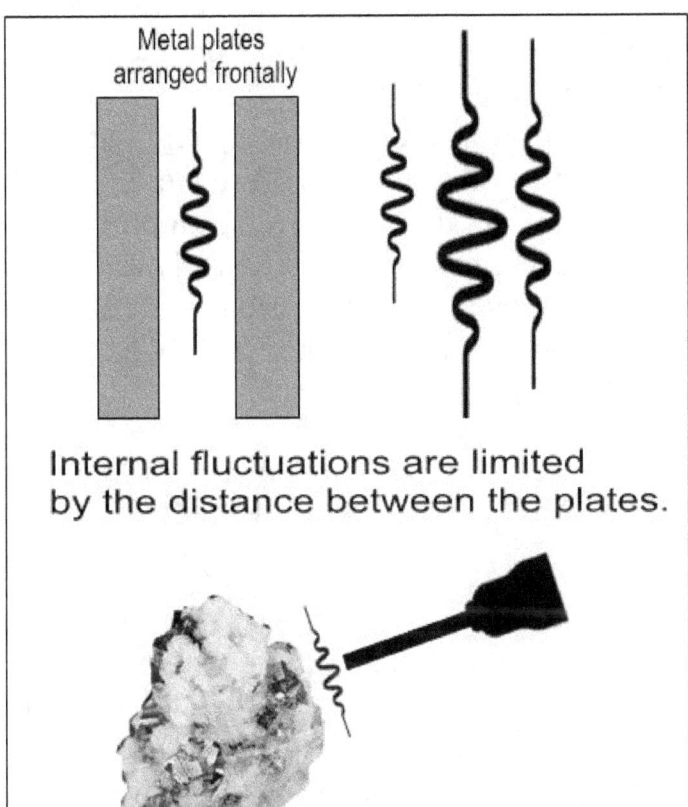

Figure 20 - The quantum atmosphere of Frank Wilczek. The materials produce a measurable aura.

Just 200 seconds after the Big Bang, helium began to form from the fusion of pairs of hydrogen atoms, and beryllium was born from the fusion of two helium atoms.

The next step was the fusion of beryllium atoms with helium atoms, and this led to the birth of the carbon atom. As this was unstable, a process was developed that made it stable. Finally, the explosion of the stars allowed the carbon to reach all the planets, and in particular the Earth, where it became the basis of life.

Every aggregation comes from a project

Why do atoms and molecules aggregate to form the body of the amoeba, fully functional in all respects? Why do other atoms aggregate to form the body of a fly, a dolphin or an elephant?

If everything were due to chance we would have had very few perfectly functional aggregations and a huge amount of irrational aggregations. But where are all these other random forms of aggregation? Where are the elephants with three legs and two proboscis? Where are the oxen covered with feathers with seven fangs?

All the aggregations of matter, from the beginning of the universe, are guided by an intelligent project that transcends matter.

The project, in fact, exists before matter and does not derive from evolution, but accompanies it. Evolution represents the free will of the creature that determines itself by agreement with the designer.

Quantum atmospheres

Some physicists, studying elementary particles, have come to a surprising conclusion. If we consider a set of protons, neutrons or electrons, in some cases their sum is greater than the sum of the single parts. There is something more than that and we don't know what it is.

Frank Wilczek, a physicist at MIT and 2004 Nobel Prize in physics, together with Qing-Dong Jian, of the University of Stockholm, recently published an amazing article on the web. In the text they claim to have probed a kind of "aura" that surrounds the materials. The two scientists called this "quantum atmosphere".

The quantum atmosphere can be measured. In this atmosphere we can detect some characteristics of the materials that were previously unknown. Wilczek explains that:

"The quantum atmosphere is a subtle area of influence around a material".

According to quantum mechanics the vacuum is not completely empty, but is full of quantum fluctuations. As mentioned in the previous chapters, particle and antiparticle pairs can arise from the quantum fluctuations of the vacuum. These couples would have been the cause of the formation of the universe.

Wilczek gives this example:
- We take two electrically charged metal plates and put them close together in a vacuum.
- Quantum fluctuations can occur between these two plates
- Obviously, these fluctuations have a wavelength smaller than the distance separating the two plates.
- Outside the plates, however, fluctuations of any wavelength may occur.
- As a result, the external energy is greater than the internal one. The two plates approach each other.

This is the Casimir effect, which is something like the quantum atmosphere (figure 19 above).

In the same way that a plate suffers a force that makes it approach the other plate, a probe and a material can record the same effect. The probe works like a second plate. In addition, the probe can measure force difference. This difference in strength, or "aura", being generated by quantum fluctuations, is called the quantum atmosphere.

Once again there are new scientific acquisitions that highlight surprising aspects of reality.

This confirms two truths:

- Quantum physics is something absolutely extraordinary.
- Our level of understanding of this branch of science is still absolutely irrelevant.

We ourselves are involved in this reality, but we do not realize it. This depends on the fact that our senses are not able to interact with quantum reality: sight, hearing, touch, taste and smell are not sized to perceive the extremely small. But we have other senses, such as intuition, intelligence and the natural tendency towards mystical and spiritual realities.

These senses, which are not related to matter, can guide us to the understanding of the secrets inherent in the deepest levels of the universe.

The theory of the quantum atmosphere foresees the existence of an external component to the matter, and this is quite surprising.

However, there are studies by well-known scientists that confirm the existence of something even more amazing.

These are studies that confirm what our spiritual senses had always understood. These studies confirm on a scientific basis the existence of a component external to man: the conscience, or the soul.

These scholars prefer to speak of "conscience" rather than "soul". Consciousness is defined as "the immediate faculty of warning, understanding, evaluating the facts that occur in the sphere of individual experience or appear in the future".

In common thought, consciousness is the moral evaluation of one's action. For example, we often say "act according to conscience".

Instead, "soul" is a word used in many religions, spiritual traditions and philosophies. In these areas the soul represents the eternal and spiritual part of a living being. Usually the soul is considered distinct from the physical body. In some cases it is thought that the soul belongs to men, but not to animals.

From the modern age the soul is progressively identified with the "mind" or "consciousness" of a human being.

Therefore, consciousness and soul should indicate the same thing. However, the term "conscience" refers to a component that man possesses independently of external inputs. This is an absolutely personal property.

Instead the term "soul" has undergone cultural conditioning matured over millennia. On the basis of these conditionings the term "soul" refers instinctively to something that is given to us in the custody of a "higher Reality". This is a temporary assignment. At the end of life we will have to return the soul, possibly improved. .

In fact, consciousness and soul can represent exactly the same thing.

This is especially true if we consider that consciousness, according to the recent studies of two very famous scientists, is something that survives the body. The body, when it dies, disappears in decomposition. Instead, the consciousness that has formed together with that body remains in the universe.

What makes us conscious?

The nature of consciousness is a great mystery that has not yet been solved. There is a vast debate going on.

The theses are mainly two.

The first thesis, which we can define materialistic, states that consciousness is only a by-product of the chemical processes that develop in brain processing. According to this thesis, consciousness could also be created with mechanical procedures, for example with a computer.

However it has been shown, on the basis of Gödel's incompleteness theorem, that our brain can perform functions that are not comparable to formal logic. No computer can reproduce such functions. Consequently, the hypothesis of a software capable of functioning as a conscience can be completely excluded.

The second thesis has a more spiritual orientation. This thesis holds that consciousness derives from some features that refer to the brain, but they have origins and destiny outside the brain. The soul is born outside man and accompanies man in existence, but does not die with man. In the universe, in addition to matter, there is a "substance" that is dispersed in the minds of living beings. That is, the universe itself is a "Spirit" or "universal Consciousness". The consciences or souls of living beings derive from this "Spirit of the universe". At the death of a being the consciousness, or soul, returns to the universal Spirit from which it comes.

Quantum physics and the soul

Two internationally renowned scientists are among the proponents of the thesis of a soul or consciousness that survives the body.

For more than twenty years a doctor and a theoretical physicist have devoted themselves to deep studies to try to understand what conscience is. Based on the latest findings of their studies the two scientists believe they are on the right path to unraveling the mystery.

One of the two scientists is Roger Penrose, a British mathematician, physicist and cosmologist (figure 19). He is known for his work in the field of mathematical physics, in particular for his contributions to cosmology. Penrose was born in 1931 in Colchester, UK, so at the time this book is published he is 87 years old.

Penrose is Professor Emeritus at the University of Oxford, he won the Wolf Prize for Physics, the Copley Medal, the Eddington Medal, the Einstein Medal and many other important honors. Furthermore, Penrose was a theorist of black holes with studies conducted together with Stephen Hawking. For his studies, Penrose was nominated for the Nobel Prize in 2008. He spent much of his life in the development of mathematical formulas capable of revealing the mysteries of the universe, including human consciousness. It is not irrelevant to note that Penrose is a convinced atheist.

In 1989 Penrose published the successful book "The emperor's new mind". In this book he states that artificial intelligence promises to give humanity a "new mind", which will be profoundly different from the mind of biological man. In the same book he argues that consciousness can originate from particular quantum phenomena that take place in brain neurons.

Recently, Roger Penrose and Stuart Hameroff published an article in "Physics of Life Reviews". In the article the authors expound new evidence in support of quantum theory on human consciousness.

Stuart Hameroff (figure 19) is an American anesthesiologist, born in Buffalo in 1947. He is currently a professor at the University of Arizona. Hameroff has developed new theories on the mechanisms that govern the functioning of human consciousness.

At the beginning of his professional career, Hameroff dedicated his studies to neoplasms and mechanisms related to the functioning of anesthetic gases. He then investigated the role played in cell division by protein structures called "microtubules".

During these studies Hameroff hypothesized that microtubules are capable of performing operations similar to mathematical calculations. Therefore, according to this scientist, microtubules have a form of "consciousness" capable of guiding and inspiring their activity.

After this finding it was easy to establish a link between understanding the phenomenon of consciousness and understanding the behavior of microtubules in brain cells. Establishing a scientific connection means starting a study.

In fact, microtubules perform functions of considerable complexity at the molecular and supramolecular level. Hameroff concludes that in cellular operations sufficient calculations can occur to speak of "conscience". In fact, the execution of a calculation involves obtaining a result and this is equivalent to a choice.

Hameroff presented these theories in his 1987 book "Ultimate Computing". The text of the book can be downloaded (in English) from the author's website at the address:

www.quantumconsciousness.org/ultimatecomputing.html

Penrose and Hameroff, combining their respective competences, have continued together the study of the phenomenon of consciousness from the point of view of microbiology and quantum physics.

The article by Penrose and Hameroff published in "Physics of Life Reviews" supports the hypothesis that consciousness is based on quantum fluctuations that occur in microtubules within brain neurons.

Moreover, these fluctuations have actually been observed and can be related to some specific electroencephalographic rhythms that had not been explained until now.

In the article Penrose disputes his critics, since all the predictions made based on his theory have been confirmed by the observations. Furthermore,

Penrose points out that his theory can be considered compatible with the two great theses present in the debate on conscience.

Penrose and Hameroff's theory is compatible with the claims of those who believe that consciousness is only a product of evolution. At the same time, the theory is compatible with the thesis of those who say that consciousness is a property of the distinct and pre-existing Universe of man..

The two researchers elaborated the "quantum theory of the mind", also called "Orch-OR". According to this theory, consciousness is a wave that vibrates in the vast subatomic universe.

The brain's microtubules act as quantum computers, receive vibrations and make them usable. In practice, microtubules produce the collapse of the probabilistic content of the conscience wave. Microtubules play the role of "observers" and transform the probabilities contained in the vibrations into well-defined "options" or "choices".

Penrose writes like this:

> "... it is quantum collapse that causes awareness. Perhaps the same collapse is awareness ".

For example. Imagine having to decide whether we should move to the right or to the left. According to some theoretical physicists, when we move to the right the alternative reality of the shift to the left collapses on the hypothesis of a shift to the right.

In the case in question, consciousness is a quantum register in which all collapses, ie all choices, are noted. This register is never deleted. The sum of the choices contained in the register is the conscience.

The two scientists define "consciousness" as the result of quantum processes that survive the body at death. This definition fits perfectly with the concept of "soul".

Consciousness has psychic and absolutely nonmaterial content. Although it is generated in the physical context of microtubules, consciousness is the psychic condensation of quantum fluctuations and collapses.

Since quantum collapses are choices, consciousness or soul would be the recording and sum of the choices made in life. Consciousness is destined to survive the body for eternity.

There is a non-secondary consequence of this theory. Every biological creature with a brain or a nervous system containing microtubules would have a soul capable of surviving forever. The soul would no longer be exclusive to man. However, while human choices may be related to free will, the choices of animals could be linked only to instinct.

Collapse of quantum waves

We increasingly talk about quantum computers. This can be the occasion to explain a practical application of the collapse of the wave function.

An ordinary computer, like the one I am using to write this text or like the one that surely each of you uses more or less often, has a memory that is now measured in billions of bits.

A Gigabit, ordinary measure of memory, corresponds to 8,589,934,592 bits. Progress has been enormous in a few decades. If someone, like me, found himself using one of the very first portable computers, the ZX80, built in 1980 by Sinclair Research by Clive Sinclair and based on the NEC μPD780C-1 microprocessor clocked at 3.25 MHz, will remember very well that his memory was 800 bits, that is 30 or 40 million times less powerful than a current tablet. Nevertheless, with the ZX80 you could do great things.

From that time, the principle underlying the operation of the memory remained the same: the data is stored in the form of bits, ie zero or one. Each computer memory is nothing but a huge deposit of these two digits, 0 and 1.

In a quantum computer, on the other hand, the memory does not contain bits but qubits. The difference is significant. In traditional computers a bit can only be 0 or only 1. Instead, in a quantum computer a qubit can be 0 and 1 at the same time. The information exists "in overlapping states", ie it functions as a probability wave. The qubits remain "undecided" in the double state of 0 and 1. When they are observed, they collapse and definitively assume one of the two possible values.

In other words, the quantum computer is able to simultaneously process many solutions to a single problem, rather than repeating the calculation many times looking for a better solution. Two qubits can have 4 states at the same time, 4 qubits have 16 states, 16 qubits have 256 states and so on.

At this time the "states" available at the same time are still few, but the research is launched towards the design of computers based on thousands of qubits. This goal would make the quantity of operations performed by a computer at the same time incalculable.

In March 2018 "Google Quantum AI Lab" presented the new 72 qubit Bristlecone processor.

Neurons as qubits

This diversity of functioning, between traditional computer and quantum computer, can be traced back to the brain. It is commonly believed that the brain works by means of interactions between neurons, as if it were a traditional computer. Each neuron can correspond to one or zero. Hameroff writes like this:

> "Most people know they own one hundred billion neurons. Consequently, they think that connections are sufficient to allow the existence of consciousness.
> These people consider the neuron as a switch that turns off or turns on, so it can be in the Zero or One state. This is an insult to the neuron itself. Just think that a single cell like paramecium swims, finds food, has the capacity to learn, finds a partner. If a simple paramecium can be so intelligent, is it possible that a neuron is so stupid? Is it just a matter of being turned on or off? I think these people don't consider what's going on inside the neuron."

Hameroff's research is focused on microtubules, absolutely complex organisms housed in neurons. Thanks to this location, the microtubules instantly respond to what happens in the mind by continually constructing and decomposing complex structures.

For example, microtubules supervise the reorganization and sorting of DNA during cell division. This is one of the most complex processes in nature. Consider the fact that any error can cause malformations.

All these considerations made Hameroff hypothesize that consciousness can be placed right inside the microtubules.

Hameroff defines microtubules as follows:

> "Microtubules are a bridge between mind and body. They transmit the wave collapse from the microscale to the human body through quantum effects, ie through that set of phenomena that occur only on a subatomic scale ".

Although Penrose lacks religious orientations, he hypothesizes with Hameroff that the quantum consciousness of every living being is independent of the body itself, and can survive the physical death of the individual.

We can say, quoting the Italian poet Silvana Stremiz:

> "There is a" sacred "place called the soul, where everything that matters is indelibly engraved. Words, gestures and thoughts are photographed for eternity ".

After death the quantum consciousness can enjoy an infinite existence, since the quantum information obeys the law of conservation of energy and therefore cannot be destroyed.

Penrose and Hameroff proposing the "theory of quantum consciousness", try to explain the experiences at the borders of death, the so-called NDE (Near Death Experience).

The two scientists monitored people near death. The observations showed that the microtubules in the brain of people close to death showed the loss of a "substance". This substance does not degrade, but disperses outside the body. In fact, in cases of "awakening", the substance returns inside the microtubules.

Angels, demons and souls of the dead

At the end of this chapter we can make some metaphysical considerations.

At present the situation is this. Although the soul is linked to a body and comes from the body itself, in effect it is not physical in nature and perhaps it does not even have a psychic nature, but has a quantum nature.

If these studies finally confirmed the existence of a consciousness or soul that survives the body, we could ask ourselves about the nature of this soul.

The soul would be the result of all quantum decision processes, that is, of all the infinite quantum collapse reactions generated by the choices made by the individual during his existence. The soul would actually be the result of all the actions done in life. All the individual's actions would be cataloged and recorded in a nebula made of quantum fluctuations.

The question that arises is this: how many of these "quantum nebulae" live around us? The answer is simple: an incalculable number.

This simple affirmation makes it clear how deep and unfathomable the mystery of souls is.

Is it possible that these very nebulae form the largest nebula that Jung called the collective unconscious? In fact, Jung speculated just that. According to the Jungian theory, the collective unconscious contains the experience of all humanity previously experienced.

This observation makes the concept of collective unconscious that could appear uniform, gray and anonymous more familiar. The collective unconscious is dressed with what is most precious, that is the memory of the people known. Not only the individuals who lived ten thousand years ago, but also the people closest to us, those we have known in our lives.

To conclude, we cannot help but formulate some hypotheses.

Perhaps even angels and demons could come out of the corner of myths to become real existences. Perhaps angels and demons are quantum condensations wanted by the universal Mind without the need to pass through a body.

And finally, a rather disturbing hypothesis. If really the souls of the dead are the aggregation of all the quantum fluctuations of their life, then they are "information". Perhaps, in a future that we do not know how close or distant it will be, could technology develop tools to intercept and decode this information?

That is, won't it be possible to communicate with the souls of the dead? Will it not be possible to dialogue with their conscience, even if it were dispersed in the non-local level of a currently unfathomable cosmos?

If this happened it would be possible to discover many truths that we had already given up. We would discover the real names of many murderers, the hiding places of many treasures, the reasons behind so many incomprehensible actions. We would discover the extreme repentance of the heroes and the extreme cowardice of the brave. We could talk to the people who were dearest to us, to tell them the words we had never been able to say during life.

There is a huge problem. Would this technology, if it could ever be realized, be morally sustainable? Would it not be right to let these souls rest in peace in their eternity? I believe that there will always be some laws of the universe, wanted by the Great Mind, that will prevent this from happening. But of course no one is able to establish what Good or Evil is, outside the narrow confines of his experience. Perhaps the concepts of Good and Evil conjugate differently or cease to exist in the immense Mind of the cosmos.

To the question if we can ever see, meet and know all the souls of the dead we can respond with the words of Buddha:

> "If the Souls of all the living beings of the Cosmos were united, God would appear there!"

To discover the mystery of the soul means to penetrate the mystery of God.

Collective unconscious and archetypes

All the studies, theories and scientific confirmations just quoted suggest that there must be a Mind of the Universe, or perhaps a Cosmic Mind that oversees and guides many universes.

Can we participate in this intelligence? And how can we go about participating?

I spoke earlier of the intelligent aggregations of matter. Now we must ask ourselves if psychic aggregations are possible.

The most credible answer has probably already been given to us in the last century, with the theories of the collective unconscious and synchronicity elaborated by Carl Jung.

Synchronicity is a widespread phenomenon that we can all witness. Synchronicity is a mysterious bond that unites two or more facts, which would normally be devoid of any connection.

This means that the "randomness" between the links taken into consideration is missing. Therefore, the phenomenon cannot be scientifically framed.

Current science is based on the principle of cause and effect: whatever happens happens because some other fact has caused it. That is, an aggregate of matter (stone, tree, person) can be the protagonist of an action that subsequently generates another action, and so on. It is a cycle that is born of the collaboration between matter and time.

Synchronicity does not need matter or time. In a synchronicity, things happen without any logical connection. Two facts absolutely disconnected from each other become synchronic when the protagonist gives them a meaning. The facts are not connected by any cause, but for the protagonist there is a very evident connection. The facts are connected, but only in the psyche of the protagonist

In his studies Jung considers the facts that go beyond the limits of statistics as synchronic, that is the facts that occur in quantities greater than what would be expected if they were simple "cases".

But where are the synchronicities born? How can it be possible that we see a forgotten person in a dream, and the next day we meet that person on the street?

Jung has theorized the existence of a collective unconscious, that is a universal and common "space" to which we are all connected.

Whenever a need, a doubt, a moment of particular suffering in our lives make our psychological guard levels lower, we open ourselves to every possible source of help. That is the moment when the collective unconscious can intervene.

It follows that the phenomena of synchronicity do not always occur. Generally these phenomena occur when we need them.

But there is also a higher connection: "messages" come from the collective unconscious to guide the whole of humanity towards higher levels of knowledge. According to Jung, the collective unconscious contains the experience of all humanity lived before us.

We are not expressing religious concepts. Even the current quantum physics, faced with the behavior of elementary particles, recognizes the existence of a "guide" the universe.

There is a psychic space, called "non-locality", where things do not happen due to a game of cause and effect.

In the subatomic level there are behaviors that seem impossible because they go beyond the conditioning of classical physics.

The strange coincidences

The strange coincidences are such common experiences that no one doubts their existence. Carl Gustav Jung talks about it with an example:

> "I find that my tram ticket has the same number as the ticket for the theater I bought a little earlier. During the same evening I receive a phone call in which someone mentions that same number. It seems to me that a casual relationship is very unlikely".

The universe is intelligent. The soul exists.

There are also less striking coincidences which, however, surprise us because we consider them almost impossible to link.

Infinite examples can be cited. We "mentally" see a friend in need, and then we verify that something unpleasant really involved that person.

We avoid taking an action because of an unpleasant feeling, and then we discover that that choice has avoided us a big trouble. We dream of a friend we had not seen for years because he lives in another city, and the next day we meet him on the street.

The official science does not believe in coincidences, because it believes that these are events that happened by chance. Actually, current science, based on materialism, believes that whatever happens, is always the consequence of something else. There can be no mysterious psychic ties between the facts that involve us.

Let's take a trivial example. A person starts from point A and walks towards point B, located around the corner. After a few steps (ie some time) that person turns the corner and only then can he see what is in point B.

It is impossible to know first what is in point B. To know it we need a body capable of seeing. The body must move, and it takes time to allow it to move. Only then will the person know what is in point B.

According to science it is not possible for the spirit to turn the corner and inform the person about what is in point B.

The coincidences can be generated by prescience, dreams, premonitions, telepathy, or other. In any case, they also bring spirit or psyche into play.

The universe is not made only of matter, but of matter and psyche that together shape our reality. If this is true then many phenomena, which would be inexplicable with the parameters of materialism, become very explainable.

Today, science finds itself uncomfortable facing the novelties of quantum physics. This physics escapes the constraints of time and space, typical of materialistic science. There are established experiments that show how very distant particles in space interact with each other simultaneously. Despite being separated by immense distances, these particles behave as if they were one.

Since these particles are not connected by any physical connection, the bond that unites them can only come from the universal psyche.

Only a psychic bond, which knows no space and time, can keep them together and can ensure that each particle is informed of what happens to the other. This is the phenomenon called "entanglement".

Synchronicity and entanglement are the basis of a new science that unifies matter and psyche. This new science will accompany humanity in a great leap towards higher levels of knowledge.

Synchronicity is a phenomenon that generates extraordinary events in our existence. Among these events we remember the presentiments, the premonitions, the dreams that come true. At other times, more frequently, synchronicity consists of a succession of facts without ties between them, which acquire a precise sense in our perception.

A synchronicity occurs every time a set of "signals" leads us to an outcome so we can say "I felt it, I expected it". It is as if something or someone wanted to warn us and give us advice on behavior.

If this happens without the knowledge necessary to process the conclusion already within us, then it is a true synchronicity.

Example: you have to leave for a trip but suddenly, for a strange feeling of discomfort, you decide not to leave anymore. Later you learn that the vehicle (train, plane or other) has suffered a serious accident with many victims.

Evidently, the preventive knowledge of the disaster cannot have been worked out in our brain. According to current science, we cannot predict the future. Jung suggests that all knowledge of the universe is contained in the collective unconscious. We all, in addition to drawing on our individual conscience that contains very limited information, we can also draw on the collective unconscious that contains all the knowledge gained from the experience of man from Adam onwards.

This practically infinite knowledge is present in the collective unconscious in the form of archetypes. The archetypes are "principles of knowledge". They can manifest themselves psychically in our consciousness through means such as dreams, premonitions, sensations, or the deciphering of significant facts that occur in our daily lives.

These significant facts are coincidences. These are facts that, individually, could be considered as belonging to the case. However, as a whole, these facts are confirmed among them until they converge into a prophecy.

We may ask why synchronicities no longer occur frequently.

Jung delved into this question. He establishes a relationship between the occurrence of synchronicities and our consent for them to occur.

According to Jung, our individual conscience manages a level of vigilance that prevents dialogue with the collective consciousness. The collective unconscious can pour its aretetypes into our individual consciousness, only if it lowers its level of vigilance. That is, the dialogue between our conscience and the collective unconscious occurs only under particular conditions. Normally, we have instinctive defenses that reject this dialogue. In his essay "Synchronicity: An Acausal Connecting Principle" Jung writes:

> "Every emotional state causes a change in consciousness. Pierre Janet has defined these modifications as "abaissement du niveau mental" (lowering of the mental level. "This means that a certain narrowing of the conscience takes place and at the same time a strengthening of the unconscious ... Consequently the conscience falls under the influence of impulses and instinctive unconscious contents ".

Consciousness naturally lowers its defense levels on the occasion of psychological traumas or complex emotional events, such as a sudden change in our standard of living, a love, a betrayal, the loss of a dear person.

Sometimes the synchronistic phenomenon precedes these events, confirming that time is only our perception and at the level of the unconscious there is no before or after.

Eastern psychology teaches lowering the levels of defense of conscience. This condition can be achieved through exercises. The contact with the mysterious dimension of the universe, the "Tao", takes place by alienating oneself.

Commenting on the "Chuang-tzu", Hans Kung writes:

> "The text speaks of" sitting and forgetting "and" fasting of the heart ". This means nothing but emptying the senses and the mind. The text says:
> - *Let your ears and eyes enter into communication with your soul. Then the gods and spirits will also come to visit you* ".

In the Western conception the Tao can be considered analogous to the Spirit of the Bible. The Biblical Spirit permeates all creation. From the Spirit descend all the illuminations that guide man, and make him capable of being the "prophet of his life". In the book of Joel God says:

> "And it shall come to pass afterward, that I will pour out my spirit upon all flesh; and your sons and your daughters shall prophesy, your old men shall dream dreams, your young men shall see visions:

And also upon the servants and upon the handmaids in those days will I pour out my spirit." *(Joel 2, 28-29).*

Take the challenge

Quantum physics is upsetting scientific thought. Many physicists among the most renowned, as they deepen their studies, are convinced of the need to integrate, in the process of understanding the universe, a force that we can define as psychic.

Without the contribution of this "psychic force" the behavior of elementary particles is no longer understandable. I mentioned many of these scientists on the previous pages.

Many of them actively testify to this belief. Some publish books, others send articles to qualified scientific magazines. Others, too, do not make their thoughts public, but begin paths of spiritual awareness.

There are dozens of physicists who have approached some Eastern philosophy, particularly Buddhism or Hinduism.

The Orientalist choice is motivated mainly by a need for intellectual freedom. Many refuse to adhere to Western religious forms, because they consider them too imbued with dogmas and precepts. This circumstance contrasts with the independence of thought, which is the greatest wealth for a scientist. For many, uncritical acceptance, that is faith, cannot overcome reason.

Evidently some scientists arrive at this choice coming from environments that tend to be atheists. Unfortunately, there are still areas of science imbued with the anticlerical fervor that followed the era of the Enlightenment.

Instead, most ordinary people do not use Kantian logic to administer their own level of faith and spirituality. Throughout the history of humanity the feeling of the existence of a "superior Entity" has always been a common heritage. In fact, there has never been a people who did not have their own list of divinities and their own religious cults. The only exceptions are some authoritarian materialistic regimes, born in the last century. Fortunately, they lasted very little.

Even in the most secularized societies, among the common population, that vague and impalpable feeling that many call "nostalgia for God" continues to be well alert and present.

In a 2012 hearing entitled "Man carries within him a mysterious desire for God" Pope Francis spoke these words:

> "The desire of God is inscribed in the heart of man, because man was created by God. God does not cease to attract

man to himself. Only in God does man find the truth and the happiness he seeks ceaselessly.

This statement may seem a provocation in the context of secularized Western culture. Many contemporaries might object that they do not feel the desire of God at all. For large sectors of society, God is no longer "the expected", "the desired". For them God is a reality that leaves one indifferent.

From this point of view, the mystery remains. Man seeks the Absolute with small and uncertain steps. However, the experience quoted by Saint Augustine, called "restless heart", is very significant. It attests to us that deep down man is a religious being ".

Meditation and prayer

This book contains many incentives to encourage the search for "God", whatever its name is and through the worship, religion or philosophy desired.

There are two main methods: meditation and prayer.

The most widely used method in oriental culture is meditation. Meditation is a practice aimed at achieving greater mastery of mental activity. Those who meditate isolate themselves from all the "background noises" of everyday life to find inner peace. The term meditation indicates the "concentration of the mind in a single point". This practice is called, more precisely, "reflective meditation".

Instead, the term "contemplation" means the "rest of the mind" in its natural state, that is, in the complete absence of thoughts. This practice is called, more precisely, "receptive meditation".

In theory, meditation is a practice of self-realization, devoid of religious purposes. In fact, it is almost always associated with spiritual or even philosophical purposes. Meditation, in different forms, is an integral part of all major religious traditions.

The first written reference to meditation, in the religious sphere, is found in the sacred Hindu scriptures of the 9th century BC, the "Upanishads". Here meditation is referred to as "dhyāna".

In yoga the practice of dhyāna favors "the experience of vision". Those who have reached a higher level can reach "enlightenment", that is, the revelation of the "omnipresent divinity".

In yoga practice it is not said that "the mind is meditating". It is said that the mind is found in dhyāna, that is, in the "state of meditation".

In the West, the main method of relating to God is prayer. Prayer can often be a repetition of pre-established formulas. In other cases, prayer arises freely and spontaneously from the soul.

God (the Holy Spirit) governs the universe and predisposes everything to happen so that man can live. The action of the Spirit extends to the needs of individuals. It is a common feeling that the Spirit is near and present in everyone. Therefore, surely the Spirit receives the prayers that are addressed to it. However, prayer should not be understood as a "request", but as a "dialogue".

One of the most beautiful choices made in the field of charismatic movements is to renounce the classic prayer of request to go to the prayer of praise. The person praying asks nothing because God knows his needs. He praises God because he exists. He thanks God because surely he prepares a happy destiny for him.

In this case prayer, as mentioned, is above all dialogue. Recognition and praise rise upwards, and at the same time serenity and consolation descend towards the person who prays.

In this dialogue we should not expect God to speak through a voice that enters the ear. All religions have always claimed that divinity speaks through "signs". In an interview, the famous Italian singer Roberto Vecchioni said:

"God sends me stronger and stronger messages. I don't understand some messages. But I have the certainty that nothing is random and that everything is caused. The beginning of things may not have been a simple "bang". The foundation of faith is that there is a reason ".

Actually, someone sends us "divine signs". Whoever he is, he assumes that we can understand them. In the Gospels we read phrases like these:

"When it is evening, you say," Good weather, the sky is red. In the morning you say: - Today it will be stormy, because the sky is dark red. So you know how to interpret the appearance of the sky. So why don't you know how to interpret the signs of the times? "
(Mt 16, 2-3).
"Look at the fig tree and all the plants. When sprouts sprout, understand for yourself that by now summer is near "
(Lk 21, 29-31).

Very often the signs come to us from the sky in the form of synchronicity, as we saw in the previous chapter.

The universe is intelligent. The soul exists.

Synchronicity is a secular concept that fits perfectly into the religious context of heavenly prophecies and communications.

The synchronicities come unexpected and should not be understood as answers to our prayers, if anything we had prayed.

Instead, the synchronicities are to be understood as messages of "Someone" who takes the floor first and wants to say something useful to us. Synchronicities are understandable especially to the person who receives them. In fact, they are effective in the personal unconscious, that is, in the most intimate part of consciousness.

Whoever gets used to recognizing synchronicities and acquires the ability to decipher them opens up a channel of privileged communication with the Spirit of the world.

The greatest difficulty in deciphering synchronicities is that they mercilessly expose what we are. In reality, synchronicities highlight our miseries and our weaknesses. We often do not recognize ourselves in these merciless photographs and conclude that those messages are not addressed to us. Too often we judge ourselves better than we are.

It is true that the universe was created for man. However, man should stand with humility before the mystery of his divinity corrupted by mortal flesh.

Perhaps in this corruption there is no guilt, only necessity. The soul needs to substantiate itself in the matter in order to exist. You must humbly undergo this step. In the Tao we can tie this maxim:

> "The wise man does not wish to prove his superiority."

The man who agrees to be clothed with humility will be taken by the hand and will be accompanied to his due glory.

Jesus reminds us of this in the discourse of the beatitudes:

> "Blessed are the humble, because the Kingdom of Heaven belongs to them"
> *(Mt 5: 3)*

Appendix 1. Hamlet

Hamlet (The Tragedy of Hamlet, Prince of Denmark), probably written between 1600 and 1602, is one of the most famous dramaturgical works in the world, translated into almost all existing languages. The famous Hamlet monologue "to be or not to be" is the most representative scene of the work and is certainly the point of arrival and a test bed for the major actors.

Almost always this part of the tragedy is cited, outside of the stages, with Hamlet holding a skull. However, this is a mistake: the skull scene is in the final part of the drama (Act V) and has nothing to do with "Being or not being", which is in the central part (Act III).

The tragedy takes place in the castle of Elsinore, in Denmark, in the medieval period.

Hamlet is often perceived as a philosophical character, with tendencies that today we could ascribe to relativism, skepticism or even existentialism.

For example, Hamlet expounds a relativist thought when, addressed to Rosencrantz, he states: "There is nothing that is good or bad, but it is the thought of man that makes things good or bad."

The idea that nothing is real except the mind of the individual draws on Greek Sophism.

The sophists argued that since everything can only be perceived through the senses, and because everyone perceives things differently, there is no absolute truth: only relative truths exist.

Characters

Hamlet: he is the protagonist of the tragedy and prince of Denmark, son of Queen Gertrude and the late King Hamlet. The king had the same name as his son.

Claudio: is the current king of Denmark, Hamlet's uncle and his antagonist; he is an ambitious politician, driven by a thirst for power and no scruples.

Gertrude: Queen of Denmark and mother of Hamlet, now married to Claudio.

Polonius: chamberlain of Elsinore, father of Laertes and Ophelia.

Ophelia: daughter of Polonius, of whom Hamlet was in love.

Laertes: son of Polonius and brother of Ophelia.

Horace: a friend of Hamlet, a fellow student at the University of Wittenberg.

Fortebraccio: Prince of Norway, whose father was killed by Hamlet's father. He wants to attack Denmark for revenge.

The ghost of the king: the specter of Hamlet's father claims to have been murdered by Claudius.

Rosencrantz and Guildenstern: two courtiers, former friends of Hamlet, who are summoned by Gertrude and Claudio to try to discover the reason for the strange behavior of Hamlet.

Voltimand and Cornelius: ambassadors.

Marcello and Bernardo: the two guards who first see the ghost of the sovereign.

Reynaldo: Polonio's servant.

The plot of the tragedy

In the sixteenth century, on the walls of the city of Elsinore, the capital of Denmark, Marcello and Bernardo speak of a ghost. Orazio also arrives, who has been called to watch over the strange phenomenon.

The specter appears shortly after midnight and Orazio immediately notices the resemblance of the ghost with King Hamlet who died recently. The ghost disappears. Orazio tells Marcello that Fortebraccio is assembling an army on the borders of Norway. With this army he wants to regain some territories. These are the lands that Fortebraccio's father lost in a duel with Hamlet. The scene moves to the royal council. Present are King Claudius, Queen Gertrude, Hamlet, Polonius, his son Laertes, the two ambassadors Cornelius and Voltimando. The theme of the meeting is the question of Fortebraccio. Those present decided to send the two ambassadors from the King of Norway to negotiate. Laertes asks King Claudius to be able to leave for France, and the king grants it to him.

Horace tells Hamlet the apparitions of a ghost resembling his father. The two decide to meet at the place of the apparitions.

The ghost appears again and asks to speak with Hamlet alone. Hamlet realizes that it is the spirit of the father. When they are alone, the ghost reveals to Hamlet that his wife Gertrude and Claudio have been betraying him for a long time.

One afternoon, while the king was sleeping in the garden, Claudio killed him by pouring a deadly henbane-based poison into his ear. At the end of the tragic story the ghost asks Hamlet to avenge him.

Returning to Horace and Marcello, Hamlet does not reveal the contents of the meeting and makes them swear not to talk to any of the apparitions.

After the terrible revelations Hamlet becomes more and more closed himself, so that Claudius and Gertrude send to call Rosencrantz and Guildenstern, two friends of Hamlet at the time of the university. Claudio asks the two to investigate the melancholy of the prince.

The two talk a long time with Hamlet and, in the name of ancient friendship, they reveal the reason for their coming. However, they try to distract the prince from his melancholy taking advantage of the arrival of a theater company.

This novelty galvanizes Hamlet, not so much for leisure, but because theatrical performance offers him the possibility of putting a plan into practice.

With his plan, Hamlet wants to solve the doubt that haunts him. He wants to be sure that the ghost is his father, and that the revelations received are true.

Rosencrantz and Guildenstern are recalled by the king to find out if they have discovered anything about the Hamlet crisis. Polonius is also present at the interview. The two cannot explain the cause of the Prince's sadness. Polonius puts forward the hypothesis that Hamlet's sadness derives from Ophelia's distance.

Hamlet arrives on the scene, so Claudio and Gertrude dismiss Rosencrantz and Guildenstern.

Then they hide with Polonio and leave only Hamlet and Ophelia on the scene.

Hamlet, however, is shocked by the revelations of the spectrum and treats poor Ofelia poorly. The girl reminds him of the old promises of love, but Hamlet advises her to become a nun.

Claudio strongly suspects that Hamlet has guessed something of his crimes, so he begins to work out a project to send him to England.

Meanwhile, Hamlet agrees with the actors of the theater company to represent a drama, "The assassination of Gonzago". This representation recalls the events narrated by the spectrum. During the play, Hamlet will observe Claudio's reactions. If the king is upset, this will mean that the phantom's charges were well founded.

The plan is successful. During the scene of the poisoning, the king abandons the theater in the throes of anger. Gertrude too is upset and invites Hamlet into his room to ask him for explanations about the reasons for that performance.

The queen agrees in advance with Polonio. Polonius will be hiding in the queen's room to report the words of the interview to the king.

Unfortunately, Hamlet, while talking with his mother, realizes that someone is secretly listening. Hamlet believes he is Claudio and kills him by shouting "a mouse, a mouse". Then take the body away to bury it quickly.

Ofelia learns the death of her father Polonius. This pain, added to the amorous disappointment for Hamlet's refusal, puts her in a state of profound madness.

Hamlet, while he is about to embark to England, meets the army of Fortebraccio which is invading Danish territory.

The soldiers tell him that the territory to which they are headed is semi-desert and useless from a strategic point of view. Fortebraccio wants to conquer those territories only for reasons of honor.

Laertes, son of Polonius and brother of Ophelia, believes that his father was killed by Claudius. He gathers an army and introduces himself to the king, accusing him of the death of his father. After a long talk, Ofelia also present, the king manages to explain to Laerte the whole truth.

Meanwhile, Horace receives a letter announcing the imminent return of Hamlet.

Then Claudio proposes to Laerte to challenge Hamlet to a duel. However, he suggests that he set up a trap. Hamlet's sword will be blunted, and Laertes sword will be dipped in a deadly poison. In addition, a cup of poisoned wine is prepared. Laerte agrees.

Ofelia, completely mad, kills herself by jumping into a lake. The scene begins with two gravediggers digging the Ophelia pit.

Hamlet wonders which noblewoman is going to be buried. When he realizes it's Ophelia, he can't help running on his coffin.

Laerte fills him with insults and challenges him to a duel to the death. The following day Hamlet is called to the king's room for the challenge.

The duel begins. The queen asks for a drink but the cup of poisoned wine is served to her. Meanwhile the duelists exchange swords several times, so that both of them injure themselves with the poisoned sword.

The tragedy ends. The first to die is Queen Gertrude. Laertes, regretting having joined Claudio's ignoble plan, reveals everything to Hamlet and dies. Hamlet, in the grip of fury, strikes Claudio with the poisoned sword. Finally, even Hamlet dies.

Glossary

A guide for the Perplexed Spiritual Testament of the German economist and philosopher E. F. Schumacher (1911-1977), putative father of the "degrowth movement".

Acausal Nexus Link between two events connected to each other but not in a causal way, that is not in such a way that the one materially influences on the other.

Alchemy Ancient esoteric philosophical system expressed through various disciplines such as chemistry, physics, astrology, metallurgy and medicine. Alchemic thought is considered by many to be the precursor of modern chemistry.

Analytical psychology Method of investigation of the deep elaborated by the Swiss analyst Carl Gustav Jung.

Anthropic principle In the physical and cosmological sphere, the anthropic principle states that scientific observations are subject to constraints due to our existence as observers.

Archetype The term is currently used to indicate, in the philosophical sphere, the pre-existing and primitive form of a thought (e.g. The Platonic idea); in analytic psychology, however, it is used by Jung and other authors to indicate innate ideas and Predetermined of the human unconscious.

Atom The atom is a structure in which matter is normally organized in the physical world. Atoms are formed by subatomic constituents such as protons, neutrons, and electrons. More atoms form molecules.

Big Bang	Cosmological model based on the idea that the universe began to expand at a very high speed in a precisely definable time of the past and that this process still continues.
Bilocation	Ability of a body to be simultaneously present in two or more different places.
Bubble Universe	*See Multiverse*
Buddhism	One of the oldest and most widespread religions in the world, originated from the teachings of the Indian itinerant ascetic Siddhārtha Gautama (VI °, V ° sec. B.C.).
Casimir effect	The attractiveness force that is exercised between two extended bodies located in the void due to the presence of the quantum field of zero point. This field originates from the energy of the vacuum determined by virtual particles that are created continuously for the effect of fluctuations.
Causality	Principle that nothing happens in the world without a decisive cause.
Cave myth	The myth of Plato's Cave is one of the most famous myths or allegories of the Athenian philosopher, told at the beginning of the book Settimo de *La Repubblica*.
Chance	Trust comforted by reasonable reasons.
Chandrasekhar limit	Non-rotating mass limit that can oppose the gravitational collapse, sustained by pressure of degeneration of electrons.
Classical physics	All scopes and models of physics that do not consider the phenomena described in Macrocosm by general relativity and microcosm by quantum mechanics.
Collective unconscious	Concept of analytical psychology coined by Carl Gustav Jung. In opposition to the personal unconscious, it is shared by all men and derives from their common ancestors.
Complex	In psychology, it is a definition used to describe a series of sentiments with uncertainties and anxieties in regard to the subject concerned and not modifiable through reasoning.
consciousness	The immediate faculty of alerting, understanding, evaluating the facts occurring in the sphere

	of individual experience or envisage in a more or less near future. In the common language, the moral assessment of one's actions.
Cosmic inflation	In cosmology, inflation is a theory that assumes that the universe, shortly after the Big Bang, has gone through an extremely rapid expansion phase.
Cultural synchronicity	An event that affects entire civilizations and millions of people.
Dark Energy	The Dark Energy is a hypothetical form of energy not directly detectable diffused homogenously in space.
Determinism	Philosophical conception of a markedly mechanistic nature, according to which every phenomenon or event of the present is necessarily determined by a phenomenon or event that happened in the past.
Discovery process	Concept developed by Swiss psychiatrist Carl Gustav Jung in the '20s. It indicates the psychic process, unique and unrepeatable, of every individual who consists in the approach of self with the selves.
Double Slit Experiment	Conceived in 1805 by Thomas Young. Represents the key to understanding quantum mechanics.
Dualism	Presence of two fundamental principles, in reciprocal relationship of complementarity or opposition.
E = MC2	The formula $E = MC2$ is the theory of relativity, on the transition between two reference systems in relative motion. E it is the Energy, m the mass of a body, c the speed of light (300000 Km/s).
Entanglement	Bond of fundamental nature existing between particles constituting a quantum system. It is also said, sometimes, quantum correlation.
Entelechy	Aristotelian term to designate the reality which attained the full degree of development.
Entropy	Measure of the disorder present in any physical system.
EPR, Paradox or experiment	An ideal experiment proposed in 1935 by Einstein, Podolsky and Rosen with the aim of demonstrating that quantum mechanics could not

	be considered a complete physical theory and that there were to be hidden, unknown variables, able to complete it.
Es	According to Sigmund Freud's psychoanalytic theory, that intrapsychoic instance that "represents the voice of nature in theSoulofman." It contains the pulsional thrusts of erotic (Eros) character, aggressive and self-destructive.
Esse EST Percipi	Motto coined by George Berkeley: *being means being perceived.*
Extrasensoriale	It is called extrasensorial perception or ESP (acronymof theEnglish expression *Extra-Sensory Perception*) Any hypothetical perception that can not be attributed to the five senses.
Fermions	So called in honor of Enrico Fermi. These are the particles that follow the Fermi-Dirac statistic and are therefore equipped with a semi-full spin (1/2, 3/2, 5/2...) .
Fibonacci Series	Succession of positive integers in which each number starting with the third is the sum of the preceding two, and the first two are by definition equal to 1. It is represented by the numbers: 1, 1, 2, 3, 5, 8, 13, 21, 34, 55 etc.
Fourth excluded	Beyond the three classical laws of physics: time, Space, causality, Jung and Pauli theorized the "fourth excluded" i.e. the synchronicity.
Fundamental interactions	In physics the fundamental interactions or fundamental forces are the forces of nature that allow us to describe physical phenomena. Four forces have been identified: gravitational interaction, electromagnetic interaction, weak nuclear interaction and strong nuclear interaction.
Golden Section	The most aesthetic relationship between the sides of a rectangle that is indicated by the number 1,6180339887.
Great Mother (archetype)	In Each of us-man or woman, it makes no difference-livesthearchetype of the Great Mother. In the psychology of Jung the Great Mother is one of the numinose powers of theunconscious, an archetype of great and ambivalent power, at the

The universe is intelligent. The soul exists. 255

	same time destructive and rescuer, nurse and devouring.
Heisenberg's Indeterminate principle	It is not possible to measure at the same time and with extreme accuracy the properties that define the state of an elementary particle. If, for example, we could determine the position with absolute precision, we would have maximum uncertainty about its speed.
Hertz	Thehertz (*symbol Hz*) istheunit of measurement of the international frequency system. It takes its name from the German physicist Heinrich Rudolf Hertz who brought important contributions to science, in the field ofelectromagnetism.
Hologram	Photographic slab or film reproducing thethree-dimensional image of an object obtained by the technique ofholography.
How much	In Quantum mechanics It is called "how much" a discrete and indivisible quantity of a certain magnitude. By extension the term is sometimes used as a synonym for "particle".
I (Ego)	In Psychology It represents a psychical-organized and relatively stable structure, connected to contact and relations with reality, both internal and external.
Idea	Term used since the dawn of philosophy, originally indicating a primordial and substantial essence. Today it has taken on the common language a more restricted meaning, generally referable to a representation or a project of the mind.
Immaterialismo	Term coined by the Irish philosopher-theologian George Berkeley (1685-1753) to define his doctrine denying theexistence of Matter.
Implicit order and explicit order	According to David Bohm in the universe there is an implicit order (implicate order), which we are unable to perceive, and an explicit order (explicate order), which we perceive.
Indeterminism	Philosophical attitude that opposes determinism.
Instinct	Internal thrust, congenital and immutable, to act and behave in a certain way. Although it is independentof intelligence, it can be modified, adjusted or repressed by the same.

Interference shape	In physics the phenomenon of interference is a phenomenon due to the overlap, in a point of space, of two or more waves.
Interpretation to many worlds	According to this theory, every time the world faces a choice at the quantum level, the universe is divided into two..
Libido	Literally translatable as desire or voluptin'. It identifies a pivotal concept of psychoanalytic theory. According to Freud it indicates the dynamic expression of the sexual impulses; According to Jung, however, the Vital and creative energy of instinct.
Lifeblood	An expression known mainly in the field of French culture, usually used in parapsicology and spiritual sciences.
Locality principle	The locality defines the area in which there are manifestations of energies and events delimited by the laws of classical physics; In the local reality there is causality or determinism (each event is determined by a previous event).
Location	In physics, the principle of locality states that distant objects cannot have instantaneous influence on one another: An object is directly influenced by its immediate vicinity.
M or Mother theory	Theory, still incomplete, that tries to mathematically combine the five theories of *superstrings* and the *supergravity to 11 dimensions* including the four *fundamental interactions*, to represent a Possible *theory* altogether.
Macrophysical Reality	The one in which we live, different from the microscopic reality relative to dimensions too small to be valued by our senses.
Mandala	Term which in particular intends to indicate an object, also sacred, of "round shape", or a "disc", especially referring to the sun or the moon. In the Buddhist and Hindu religious tradition, symbolic representation of the cosmos, made with threads woven on the frame or with powders of various colours on the ground, or painted on cloth, or frescoed on the walls of the temple.

Manichaeism	Radically dualistic religion: two principles, light and darkness, independent and contrasting affect every aspect of human existence and conduct.
Mass number	Indicates the number of nucleons (i.e. protons and neutrons) present in an atom.
Matter	In classical physics, with the term materia one indicates generically anything that has mass and occupies space; or alternatively, the substance of which the physical objects are composed, thus excluding the energy, which is due to the contribution of the fields of force.
Mentalism	Philosophical conception that tends to reduce knowledge data to pure perceptions of the mind, neglecting the objective aspects of physical experience.
Metaphysics	Philosophical doctrine that presents itself as the science of absolute reality and which seeks to give an explanation of the first causes of reality, regardless of any given of the experience.
Mind uploading	Recovery and transfer of the mental patrimony of an individual from the old to a new body.
Monism	Monism is a conception of being that is opposed to that of pluralism, or more often to that of dualism.
Multiverse	A parallel dimension or parallel universe is a hypothetical universe separate and distinct from our but coexistent with it; In most of the cases imagined it can be identified with another space-time continuum. The whole of any parallel universes is called Multiverse.
Mystery religions	The main mystery cults. The most famous mysteries of the Greek world were the *eleusine mysteries*, linked to the cult of Demeter and Persephone. Beside these are to remember those related to the cult of Dionysus and Orpheus in the *orphic Mysteries* and the cult of the Phrygian god Sabazio; Finally, the *Mysteries of the Cabirs* in Samothrace.
Myth	The myth is a story clad in sacredness. It is a story about the origins of the world or the ways in which the world itself or living creatures have reached the present form.

Mythology	Thewhole of the fantastic or religious elaborations of a certain cultural tradition.
Neoplatonism	Interpretation of Plato's thought given in the Hellenistic age. To summarize in itself several other elements of Greek philosophy, and become the main ancient philosophical school from the THIRD century.
Neurosis	Mental disorder of a predominantly psychological nature, derived from an unconscious conflict betweentheindividual and theenvironment.
Newtonian physics	*V. Classical physics*
Nigredo	In alchemy the phase with the black of the great work, that of rot and decomposition, that is the initial step in the path of creation of the Philosopher's Stone.
Nirvana	A concept that indicates a state of happiness, precisely of the Buddhist and Jain religions, later introduced in Hinduism aswell.
Nobel Prize	A world-value award attributed annually to people who have distinguished themselves in the various fields of knowledge, "bringing greater benefits to the allumanity" for their researches, discoveries and inventions, fortheliterary work, Forthecommitment to world peace.
Non-Location	The level in which the physical principles of the locality are no longer valid.
Notarikon	Hebrew method to derive a word, in a manner similar to the creation of an acronym, making sure that each of its initial or final letters representanotherword.
Numinous	Surrounded by a halo of sacredness that enchers with fright and reverence.
Objectivity	Ideological representation corresponding to reality, to the objective world, and thus not dependent on anactivity of conscience.
Old Sage (archetype)	Embodiment of the spiritual principle. Usually theindividual encounters such an archetype in critical situations of his own life when he has to make difficult decisions.

Olomovimento	Term coined by Bohm to describe theuniverse as a dynamic system in continuous motion. Instead, the term hologram usually refers to a static image.
Omega Point	Term coined by the Jesuit scientist French Pierre Teilhard de Chardin to describe the highest level of complexity and consciousness to which it seems that theuniverse tends to evolve.
Oort Cloud	The Oort cloud is a spherical cloud of comets placed at a distance from the Earth equal to about 2400 times the distance between the sun and Pluto.
Opus alchemicum	Alchemic procedure to obtain the Philosopher's stone, which occurred through seven procedures, divided into four operations: rot, calcination, distillation and sublimation, plus three phases: solution, coagulation and dyeing.
Orbital	Wave functiondescribing the behavior of an electron in an atom.
Paleolithic	Period characterised by the construction anduse of stone tools with more and more refined workings, and which sees the beginning, inMan, of metaphysical thought and of the cult of the dead.
Parallel size	A parallel dimension or parallel universe is a hypothetical separate universe and distinct from ours but coexistent with it.
Parallel universes	A parallel dimension or parallel universe is a hypothetical universe separate and distinct from our but coexistent with it; In most of the cases imagined it can be identified with another space-time continuum. Thewhole of any parallel universes is called Multiverse.
Paranormal	Term that applies to phenomena that are contrary to the laws of physics and scientific assumptions.
Periodic table of the elements	A scheme by which the chemical elements are sorted on the basis of their atomic number Z and the number of electrons present in the atomic orbitals.
Person (archetype)	One of the Jungian archetypes that derives its name from the Latin, where it has the meaning of "mask of theactor" and indicates the role that the subject interprets in the social context in which it acts.

Photon	The photon is the "quantum" of the electromagnetic field, historically also called the quantum of light.
Pilot Wave	Interpretation of quantum mechanics postated by David Bohm in 1952. It incorporates theidea ofthepilot wave elaborated by Louis de Broglie in 1927.
Planck's constant	Physical constant representing theminimum action possible. It determines thatThe associated fundamental physical energy and quantities do not evolve continuously, but are quantized.
Principle of complementarity	In quantum mechanics, it states that the dual aspect (wave and particle) of some physical representations of atomic and subatomic phenomena cannot be observed at the same time during the same experiment.
Principle of non-locality	Principle of quantum mechanics according to which subatomic particles are able to communicate information instantaneously.
Probabilism	Intermediate doctrine between dogmatism and skepticism, alleging that the objectively safe knowledge of reality is not possible.
Psychoanalysis	Term that derives from *Psycho*, Psyche, soul, and *analysis*: Analysis of the mind. It is the theory of theunconscious ofthehuman soul on which a discipline is founded, known as Psychodynamics, and a related psychotherapeutic practice, which tookthestart from the work of Sigmund Freud, who He inserted in the groove of the works of Jean-Martin Charcot and Pierre Janet.
Psychology of form	The psychology of Gestalt (from the German *Gestaltpsychologie*, *Psychology of the form* or *representation*) is a psychological current centered on the themes of the perception andof theexperience .
Quantum Leap	Instantaneous change of a system, which occurs on a very small scale and is randomly held. For example, an electron that, being in an energy level of an atom, jumps instantaneously into a different energy level.

Quantum mechanics	Physical theory describing the behavior of matter, of radiation and of reciprocal interactions, with particular regard to the characteristic phenomena of the scale of length or of atomic and subatomic energy.
Quantum physics	Physical theory describing the behavior of matter and radiation and the reciprocal interactions, with particular regard to the characteristic phenomena of the subatomic level of magnitude.
Quantum potential	Parameter added by David Bohm totheSchrödinger equation. Quantum potential transforms quantum mechanics from probabilistic theory to deterministic theory.
Quantum theory	V. *Quantum mechanics*.
Quark	In physics, the fundamental constituents of hadronic matter, that is, of all observed particles that are subject to strong interactions.
Reductionism	Reductionism in general maintains that the institutions, methodologies or concepts of a science must be reduced to the lowest common denominators or to the most elementary entities possible.
Reincarnation	Cyclical resurrection that culminates with the attainment of perfection.
Republic	The *Republic* is aphilosophical work in the form of a dialogue that has had enormous influence in Western thought, written roughly between 390 and 360 B.C. by the Greek philosopher Plato.
Res Cogitans e Res extensa	With *Res Cogitans* we mean the psychic reality to which Descartes attributes the following qualities: inextension, freedom and awareness. The *res extensa* is instead the physical reality, which is extended, limited and unconscious.
Resurrection	Return to life after death, with ananalogy to awakening after sleep. Common to all religions that foresee the reviviscence of thesoul of the deceased.
Rubedo	The last phase of the great work, the "Red One". It is the final fulfillment of the chemical transmutations, culminating in the realization of the Philosopher's Stone and the conversion of the vile metals into gold.

Samsara	In the religionsofIndia such as Brahmanism, Buddhism, Jainism andHinduism, it indicates the doctrine inherent in the cycle of life, death and rebirth.
Self	Nucleus of the personality, indicated with the pronoun of a third person singular todistinguish it from the*ego*, that is, from its reflected image in which consciousness normally identifies itself.
Shadow (Archetype)	Powerful archetype, container of all that we have missed in the good and of all that we have received in evil. It is therefore our enemy, theantagonist, what appears in fairy tales as the "villain" and which is often depicted in the form of a monster, dragon or demon.
Shamanism, Shaman	With the term shamanism it is indicated, in the history of the religions, in cultural anthropology and ethnology, a set of beliefs, religious practices, magic-rituals or ecstatic techniques found in various cultures and traditions.
Sofisti	Masters of Contemporary virtues of Socrates and Plato, who were charged to donate their teachings
Soul and Animus	Archetypes with high dual content. Each archetype contains one aspect of life and its opposite, suggesting that both have their own value. Theimage ofthesoul is projected by men on women, while in women is thecorresponding archetype, theAnimus, to be screened on men.
Soul of the World	Also known in Latin as *Anima Mundi*, It is a philosophical term used by the Platonic to indicate the vitality of nature in its entirety, assimilated to a single living organism.
Spacetime	Spacetime is the four-dimensional structure of the universe. The concept of space-time has been introduced by special relativity, and is composed of four dimensions: the three of space and time.
Spin	In quantum mechanics, the spin (literally "whirling gyrus" in English) is a magnitude, or quantum number, associated with the particles that it contributes to define the quantum state. The spin is a form of angular momentum.

String theory	Theory, still in development, that tries to reconcile quantum mechanics with general relativity and that hopefully can constitute a theory altogether.
Subatomic, subatomic level	Level of the elementary particles, below the size of the atom.
Subjectivity	Personal vision of values, of a judgement, of a critique.
Super-Ego	According to Freud it indicates one of the three instances which, together with 'Es and I, compose the structural model of thepsychic apparatus.It originates from the internalization of codes of conduct, prohibitions, injunctions, schemes of value (good/evil; right/wrong; good/bad; agreeable/unpleasant) that the child implements all' In the relationship with the parents ' couple.
Supernova	A supernova is a stellar explosion. Supernovae are very bright and cause radiation emissions that can exceed those of an entire Galaxy.
Symbol	The symbol is an element of communication, expressing content of ideal significance of which it becomes the signer. Normally the symbol is something that is in the place of something else.
Synchronicity	Concept theorised by psychoanalyst Carl Gustav Jung in 1950, defined as "a principle of acausal links". It consists in a link between two events connected to each other but not in a causal way, that is not insuch a way that theone may have materially influencedon theother.
Taoism	A set of philosophical and mystical doctrines formulated by Chinese thinkers in the ESA. IV ° and III ° B.C.
Temurah	Method used by the kabalists to tidy up the Hebrew Bible's words and phrases to derive the esoteric substrate and spiritual significance.
Tetraktys	The Tetraktys or "quaternary number" represented for the Pythagoreans the arithmetic succession of the first four natural numbers, or positive integers.
The principle of overlapping states	The principle states that, just like the waves of classical physics, two or more quantum states can

	be summed ("overlapping"), and the result will be another valid quantum state.
Theology	Study of nature, oftheessence, of the attributes and manifestations of God.
Theory of relativity	The theory of relativity formulated by Albert Einstein, first in its narrow version and then in the general one, profoundly altered the theory of Galilean relativity and changed our concept of time and space. However surprising, Einstein's predictions have achieved numerous confirmations.
Theosophical Reduction	Method whereby all numbers can be reduced to a single digit between 1 and 9.
Thesubatomic level	Level in which you have smaller dimensions than those of the atom, or which relates to the constituent parts of the atom, such as electrons, neutrons etc.
Time Arrow	Phenomenon according to which time seems to flow always in the same direction, from the past to the future, according to a kind of unique sense. It defines the time arrow the phenomenon (real, observable and complex) such that a physical system evolves from an initial state S in time T to a final state S2 at a time T2 and will never return to the state S.
Transcendent	Not attributable to the determinations ofthe experience, as it subsists independently of the reality of which it is also the assumption.
Transhumanism	Cultural movement that advocates theuse of science and technology to increase man's physical and mental capacities.
Unconscious	All mental activities that are not present to the conscience of an individual.
Upanishad	In Sanskrit, "arcane Doctrines, secret". Denomination of a series of philosophical-religious textsofIndia, belongingtothe last phase of the Vedic period.
Vitalism	A current of thought that Plato's ideas must be extended to all nature, and become a constituent part of every single organism and of all that exists.

The universe is intelligent. The soul exists.

Wave function	In Quantum mechanics the wave functionrepresents the state of a physical system. It is a complex function of spatial and time coordinates and its meaning is that of anamplitude of probability.
Wheel of Medicine	In the culture of the American Indians the wheel of medicine is traditionally built with stones or sticks based on the four sacred directions of space.
World of Ideas	TheHyperuranium, or world of ideas, is a concept of Plato expressed in *Phaedrus*.
Wormhole	Shortcut between two points of the universe.

Bibliography

Amir Dan Aczel, Entanglement. The greatest mystery of physics.
Barbour Julian, End of the time.
Barrow John David, From zero to infinity. The great story of Nothing.
Barrow John David, The numbers of the universe,
Barrow John David, Why is the world a mathematician?
Barrow John David, look Frank The anthropic principle.
Beitman Bernard, Messages from coincidences.
Cambray Joseph, Synchronicity. Nature and Psyche In a connected universe.
Cantalupi Tiziano, Santarcangelo Donato, Psychism and reality. .
Capra Fritjof, The Tao of physics.
John Cederquist, Coincidences They don't exist.
Cesati Cassin Marco, We're not here by chance.. The power of coincidences.
Subrahmanyan Chandrasekhar, Truth and Beauty. The reasons for aesthetics in science.
Chinnici Giorgio, Case Guard. The secret mechanisms of the quantum world
Chopra Deepak, Coincidences
Ford Kenneth, The world of Quanta. Quantum physics For everyone.
Gamow George, The Adventures of Mr. Tompkins.
Gamow George, Mr. Tompkins ' New World.
Goswami Arneb, Quantum Lighting Guide.

Greene Brian, The plot of the cosmos. Space,
Greene Brian, The hidden universes of parallel reality And the profound laws of the cosmos.
Greene Brian, The elegant universe. Superstrings, hidden dimensions and the pursuit of definitive theory.
Hawking Stephen The Universe in a nutshell.
Hawking Stephen The theory completely. Origin and destination Dell Universe.
Hawking Stephen The great history of the time.
Hawking Stephen Do Big Bang For black holes. A brief history of the universe.
Heckler, Richard, Coincidences.
Robert Hopke, Nothing happens by chance.
Joseph Frank, The power of coincidences.
Young Carl The analysis of Dreams. Archetypes of the unconscious. Synchronicity.
Young Carl Memories, DreamsReflections.
Kane Gordon, The Garden of Particles Elemental.
Shani Mani Quantum. From Einstein In Bohr, quantum theory, a new idea of reality..
Rei Hans, Christianity and Chinese religiosity.
Lederman Leon, Hill Christopher, Physical Quantum for Poets
Licata Ignazio, Watching the Sphinx.
Motterlini Matteo, Mental traps.
Peat David, Synchronicity. A union between the matter e Psyche.
Popper Karl, The Ego and your brain.
Radin Dean. Intertwined minds. Psychic phenomena explained by quantum physics.
Rhine Louisa, Psychokinesis. in mind Dominates matter..
Schumacher Ernst, A guide to the Perplexed, the B
Sheldrake Rupert, The illusions of Science.
Sheldrake Rupert, The mind Extended..
Michael Smith, Young and Shamanism.
Sparzani and Panepucci. (Curators) Young and Pauli. The original correspondence: The meeting between psyche and matter.
Henry Stapp Quantum theory and free will..
Michael Talbot, All is a. Feltrinelli
Teodorani Massimo, Bohm. The Physics of Infinity.
Teodorani Massimo, in mind Creative. From the physical universe to intelligent life.

The universe is intelligent. The soul exists.

Teodorani Massimo, The entanglement. The Weave In the quantum world: particles To consciousness.
Teodorani Massimo, Synchronicity. The link between physics and psyche. Da Pauli Young ' s Next In Chopra.
Teodorani Massimo, The Atom and the particles Elementary.
Seems Frank The physics of Immortality.
John White, The encounter between science and spirit..
Claudio Widmann, Synchronicity and coincidences Significant.
Claudio Widmann, Introduction to Synchronicity.

www.ingramcontent.com/pod-product-compliance
Lightning Source LLC
Chambersburg PA
CBHW070618220526
45466CB00001B/50